AF838315

UNIVERSITY OF SUNDERLAN
INFORMATION SERVICES

"LEASE RETURN TO: THE MU

Osmoregulation and Drinking in Vertebrates

UNIVERSITY OF SUNDERLAND
WITHDRAWN

EXPERIMENTAL BIOLOGY REVIEWS

Series Advisors

D.W. Lawlor
AFRC Institute of Arable Crops Research, Rothamsted Experimental Station, Harpenden, Hertfordshire AL5 2JQ, UK

M. Thorndyke
School of Biological Sciences, Royal Holloway, University of London, Egham, Surrey Tw20 0EX, UK

Environmental Stress and Gene Regulation
Sex Determination in Plants
Plant Carbohydrate Biochemistry
Programmed Cell Death in Animals and Plants
Biomechanics in Animal Behaviour
Cell and Molecular Biology of Wood Formation
Molecular Mechanisms of Metabolic Arrest
Environment and Animal Development: genes, life histories and plasticity
Brain Stem Cells – SEB Symposium Series Vol. 53
Endocrine Interactions of Insect Parasites and Pathogens
Osmoregulation and Drinking in Vertebrates

Forthcoming titles include

Vertebrate Biomechanics and Evolution Vols I and II

Osmoregulation and Drinking in Vertebrates

Edited by

N. HAZON
School of Biology, University of St Andrews, Gatty Marine Laboratory, St Andrews, Fife, KY16 8LB, Scotland, UK

and

G. FLIK
Department of Animal Physiology, Faculty of Science, University of Nijmegen, Toernooiveld 1 6525 ED, Nijmegen, The Netherlands

UNIVERSITY OF
SUNDERLAND

UNIVERSITY LIBRARY

BIOS

© BIOS Scientific Publishers Limited, 2002

First published 2002

All rights reserved. No part of this book may be reproduced or transmitted, in any form or by any means, without permission.

A CIP catalogue record for this book is available from the British Library.

ISBN 1 85996 094 4

BIOS Scientific Publishers Ltd
9 Newtec Place, Magdalen Road, Oxford OX4 1RE, UK
Tel. +44 (0) 1865 726286. Fax + 44 (0)1865 246823
World Wide Web home page: http://www.bios.co.uk/

Distributed exclusively in the United States, its dependent territories, Canada, Mexico, Central and South America, and the Carribean by Springer-Verlag New York Inc., 175 Fifth Avenue, New York, USA, by arrangement with BIOS Scientific Publishers Ltd., 9 Newtec Place, Magdalen Road, Oxford OX4 1RE, UK

UNIVERSITY OF SUNDERLAND LIBRARY
Class No. 596.0132 HØ9
Access No. 1111371085
Order No. 90

Production Editor: Aimie Haylings
Typeset by Kolam Information Services Pvt. Ltd., Pondicherry, India
Printed by The Cromwell Press, Trowbridge, UK

Contents

Contributors

Acher, R., Laboratory of Biological Chemistry, University of Paris VI, 96 Boulevard Raspail, 75006 Paris, France

Anderson, W.G., Gatty Marine Laboratory, School of Biology, Division of Environmental and Evolutionary Biology, University of St Andrews, Fife KY16 8LB, UK

Balment, R.J., School of Biological Sciences, University of Manchester, Manchester M13 9PT, UK

Chauvet, J., Laboratory of Biological Chemistry, University of Paris VI, 96 Boulevard Raspail, 75006 Paris, France

Cramb, G., Bute Building, School of Biology, University of St Andrews, St Andrews, Fife KY16 9TS, UK

Cutler, C.P., Bute Building, School of Biology, University of St Andrews, St Andrews, Fife KY16 9TS, UK

Fan, Z., University of Tennessee, Department of Physiology and Biophysics, 894 Union Ave., Nash Building, Room 550, Memphis, TN 38163, USA

Fenwick, J.C., Department of Biology, University of Ottawa, P.O. Box 450 STN A, Ottawa, Ontario K1N 6N5, Canada

Flik, G., Department of Animal Physiology, Faculty of Science, University of Nijmegen, Toernooiveld 1 6525 ED Nijmegen, The Netherlands

Fuentes, X., Centro de Ciencias do Mar, Campus de Gambelas, Estrada de Penha, Universidade do Algarve 8000–810, Faro, Portugal

Grau, E.G., Hawaii Institute of Marine Biology, University of Hawaii, PO Box 1346, Coconut Island, Kaneohe, Hawaii 96744, USA

Guerreiro, P.M.G., Department of Animal Physiology, Faculty of Science, University of Nijmegen, Toernooiveld 1 6525 ED Nijmegen, The Netherlands

Hazon, N., Gatty Marine Laboratory, School of Biology, Division of Environmental and Evolutionary Biology, University of St Andrews, Fife KY16 8LB, UK

Hirano, T., Hawaii Institute of Marine Biology, University of Hawaii, PO Box 1346, Coconut Island, Kaneohe, Hawaii 96744, USA

Huising, M.O., Department of Animal Physiology, Faculty of Science, University of Nijmegen, Toernooiveld 1 6525 ED Nijmegen, The Netherlands

Leedom, T.A., Hawaii Institute of Marine Biology, University of Hawaii, PO Box 1346, Coconut Island, Kaneohe, Hawaii 96744, USA

Lignot, J-H., Université Louis Pasteur, CNRS CEPE, 23 Rue Becquerel, 67087 Strasbourg, Cedex 2, France

Nishimura, H., University of Tennessee, Department of Physiology and Biophysics, 894 Union Ave. Nash Building, Room 301, Memphis, TN 38163, USA

Rankin, J.C., Institute of Biology, Odense University, Campusvej 55, Odense M, Denmark

Seale, A.P., Hawaii Institute of Marine Biology, University of Hawaii, PO Box 1346, Coconut Island, Kaneohe, Hawaii 96744, USA

Takei, Y., Laboratory of Physiology, Ocean Research Institute, University of Tokyo, 1-15-1 Minamidai Nabano 164–8639, Tokyo, Japan

Uchiyama, M., Department of Biology, Faculty of Science, Toyama University, Gofuku 3190, Toyama 930-8555, Japan

Varsamos, S., Department of Animal Physiology, Faculty of Science, University of Nijmegen, Toernooiveld 1 6525 ED Nijmegen, The Netherlands

Warne, J., School of Biological Sciences, University of Manchester, G38 Stopford Building, Oxford Road, Manchester M13 9PT, UK

Wells, A., Gatty Marine Laboratory, School of Biology, Division of Environmental and Evolutionary Biology, University of St Andrews, Fife KY16 8LB, UK

Yoshizawa, H., Department of Biology, Faculty of Dentistry, Matsumoto Dental University, Gouhara 1780, Shiojiri 399-0781, Japan

Abbreviations

BP	Before present
^{51}Cr-EDTA	Radiolabelled solution
FW	Freshwater
i.p.	Intraperitoneal
i.v.	Intravenous
MS222	3-aminobenzoic acid ethyl ester methane sulphonate salt
p.p.t.	Parts per thousand
SEM	Standard error of mean
SW	Sea water
ABC	Avidin–biotin–horseradish peroxidase complex
ACE	Angiotensin converting enzyme
ADH	Antidiuretic hormone
ANG I	Angiotensin I
ANG II	Angiotensin II
ANP	Atrial (A-type) natriuretic peptide
AQP	Aquaporin
AQP-TB	Aquaporin-toad bladder
AV3V	Antero-ventral third ventricle
AVP	Arginine vasopressin
AVT	Arginine vasotocin
BK	bradykinin
BNP	Brain (B-type) natriuretic peptide
cAMP	Cyclic adenosine monophosphate
CD	Collecting duct
CHIP	Channel-forming integral protein
CNP	C-type natriuretic peptide
C-Rfa	*Carassius* RFamide
CSF	Cerebrospinal fluid
DL	Descending limb
DR	Drinking rate
EDTA	Ethylenediamine tetraacetic acid
EnaC	Epithelial sodium channels
FA-CHIP	Frog AQP-CHIP
fMIP	Frog lens MIP
FSK	Forskolin
FW	Fresh water or freshwater
GFR	Glomerular filtration rate
GLP	Glycerol facilitator-like proteins

HPLC	High pressure liquid chromatography
ICV	Intracerebroventricular
IMCD	Inner medullary collecting ducts
IMPs	Intramembranous particles
i.p.	Intra-peritoneal
JG	Juxtaglomerular
KKS	Kallikrein–kinin system
MIPs	Major intrinsic proteins
MRCs	Mitochondria-rich cells
MS222	3-aminobenzoic acid ethyl ester methane sulphonate salt
NDI	Nephrogenic diabetes insipidus
NP	Natriuretic peptide
NPA	Asparagine-proline-alanine
NPR-B	CNP-specific receptors
NPS	Natriuretic peptide system
PAM	Polypeptidylglycine-alpha-amidating enzyme
PAP	Peroxidase–antiperoxidase
PC	Principal cells
PKA	Protein kinase A
PLC-IP3	Phospholipase C-phosphatidylinositol
PNA	Peanut agglutinin
PRL	Prolactin
ProVT	Pro-vasotocin
PrRP	PRL-releasing peptide
PTHrP	Parathyroid hormone related protein
PVN	Paraventricular nuclei
RAS	Renin–angiotensin system
RIA	radioimmunoassay
RT-PCR	Reverse transcription-polymerase chain reaction
SFO	Subfornical organ
SMIT	Sodium/*myo*-inositol-cotransporter
SON	Supraoptic nuclei
SW	Sea-water
TAL	Thick ascending limb
TL	Thick limb
TMAO	Trimethylamine oxide
TonE	Tonicity-response enhancers
TonEBP	TonE binding protein
U/P	Urine-to-plasma
UT	Urea transporter
VAMP2	synaptobrevin
VNP	Ventricular natriuretic peptide
W	Weight

Preface

The purpose of this volume is to provide a comprehensive overview of the comparative physiology of drinking and water balance in vertebrates. The chapters were originally presented as papers at a meeting organized by the Society for Experimental Biology in Sorrento, Italy (26th–30th May, 2001)

The first chapter deals with lampreys. The extant species are representatives of the earliest vertebrates; their metamorphosis from larva to adult reflects the transition from invertebrates to vertebrates. In freshwater they drink minimally, in seawater copiously to compensate for osmotic water efflux. Aquaporin (AQP) analogues may be absent in the gills of these animals, indicated by similar osmotic and diffusional water permeabilities of the branchial epithelium. The involvement of the renin–angiotensin system in the regulation of drinking behaviour in response to variations in ambient salinity is still doubtful.

In chapter 2 elasmobranch species (next in phylogeny to the hagfishes and lampreys) appear on stage. Marine sharks counteract osmotic water loss by urea retention (the blood plasma is slightly hypertonic to the ambient medium), and thus drinking is not really a requirement. Yet the mechanisms are there: it is shown that drinking is induced in response to a suddenly increased ambient salinity and, most importantly, in response to hypovolemia. The renin–angiotensin system plays a central role in dipsogenesis.

The next 5 chapters present the state of the art in research on fishes. Fish, with their estimated 35 000 extant species, apparently are the most successful vertebrates and, indeed, have developed strategies to live in media that range from almost distilled water to concentrated seawater. The extensive surface of the delicate gills and their permeability to water require that fish drink in hypertonic media to counteract osmotic water loss. They so do. But also in fresh water fish drink, maybe to moisten their food and certainly when stressors activate the endocrine system(s) to produce cortisol. In chapter 3 it is shown that very young fish start drinking as soon as they have shed their egg membrane (which is notoriously impermeable to water), in hypertonic media and when stressed. Stress-induced drinking may be used as a tool to enhance vaccination efficiency and may enhance exposure to waterborne toxicants (via leaky intestinal routes). In chapter 4 we follow the seawater drunk by eels to provide the fish with water and its processing to avoid intoxication with the high levels of magnesium in seawater. An array of ion and water transporters, from ATPases, cotransporters and exchangers to water channels, elucidated by molecular biological techniques, provides definitive substrate for previously postulated models for water handling by the fish intestine (chapter 4). In chapter 5 the hormonal control of drinking in the Japanese eel is addressed. A multitude of internal (osmotic and volemic stimuli, angiotensin-II and the heart peptide ANP) and external (water

Cl^-) signals determine the dipsogenic reponse. The kallikrein–kinin system antagonizes the natriuretic peptide surge (which by itself acts antidipsogenically) after seawater transfer and explains the drinking seen upon seawater transfer. In chapter 6 the intricate role of arginine vasotocin (AVT) in fish osmoregulation is addressed. AVT at physiological concentrations has antidiuretic effects in seawater fish, consistent with a role in seawater residence and adaptation to this medium. In chapter 7 it is shown that taking blood from tilapia (an advanced teleostean species) induces drinking to restore plasma volume at the expense of plasma osmolarity (going down in fresh water, going up in seawater). The decreased plasma osmolarity and the angiotensin-II surge in fresh water stimulate prolactin release and the hormone restores plasma osmolarity through a plethora of actions. In seawater cortisol and other hormones are crucial in the restoration of plasma osmolarity.

World-wide, very little research has been done on amphibian species, but in chapter 8 we learn that frogs are able to drink through their abdominal skin if necessary and are the first to have developed a specialized, functionally defined (distal) nephron to cope with terrestrial conditions of temporary limited water availability. There is reference to permanently aquatic and seawater crab-eating frogs!

In chapter 9, dealing with birds, the conservation of water is the issue and here the kidney is crucial. The avian nephron comes in two sorts (reflecting phylogenetic stages?), with and without analogues of the loop of Henle, where urine is either concentrated by addition of NaCl or diluted by specific NaCl reabsorption. Medullary collecting ducts (CD) are important in the definition of the eventual urine and AVT is targeted at one of at least three AQPs (*viz.* the AQP2 homologue, not the AQP1 and AQP4 homologues) to regulate water movement.

The final two chapters provide insights on drinking and fluid balance from studies on mammals. The metanephric kidney of mammals concentrates urine to the ultimate if required, and arginine vasopressin is dominant in the regulation of kidney function. As a result of the water conserving capacity of this type of kidney, drinking plays an important yet less prime role in water balance. Brain control – via sensing of derived levels of (anti-)dipsogenic hormones in cerebrospinal fluid – over drinking and fluid balance is evident. Is the system basically different from what we know of the elegant studies on the Japanese eel? The reader is invited to draw her or his own conclusions. In the final chapter of this book amphibia are revisited. Their lifestages always comprise an aqueous (too much water) and often a terrestrial phase (too little water) and it appears that the water-to-land transition has resulted in the emergence of two functionally different hormones (vasotocin and hydrin2) derived from a common precursor, showing that opposite physiological demands may depend on redirecting and readjusting a signalling machinery to functionally different effector proteins (such as aquaporins).

Special thanks go to the contributors to this volume, who handed in their original work so promptly. Significant financial support was received from the SEB, London, and the Osmoregulation Group of the SEB which we gratefully acknowledge. Finally, we want to thank BIOS Scientific Publishers Ltd, and Christine Trimmer and Mike Lakin for running the office in London and Sorrento so smoothly, without their invaluable help this book would probably not exist.

Gert Flik
Neil Hazon

1

Drinking in hagfishes and lampreys

J.C. Rankin

1. The evolution of vertebrate osmoregulation

Hagfishes and lampreys (Agnatha, or jawless 'fish') occupy an interesting position in vertebrate phylogeny. They were placed together in the class Cyclostomata but, despite their superficial resemblances, are now thought to have diverged in the Pre-Cambrian (Janvier, 1999). The earliest known vertebrate fossils from the Lower Cambrian (> 500 million years BP) in Chengjian, China, are of two species, one related to the lampreys and the other intermediate in character between the lampreys and the hagfishes (Shu *et al.*, 1999). Around 100 species of Agnatha have survived to the present day, competing with jawed vertebrates (Gnathostomata) by exploiting specialized niches. Hagfish are confined to the marine environment where they feed on invertebrates and by scavenging dead invertebrates and vertebrates (Martini, 1998). New species of hagfish are regularly described but the 38 species of lampreys are well known. Most are confined to freshwater (FW), although a few are anadromous with the filter feeding larval stage (ammocoete) in FW. The ammocoetes metamorphose into parasitic adults which migrate to the sea (or lakes) and grow to full size, feeding by sucking the blood of other fish before returning to rivers to spawn and die (Hardisty, 1979).

The vertebrates evolved from marine invertebrates but, at some stage, the ancestors of the lampreys and Gnathostomata moved from sea water (SW) into FW. Here they faced an enormous problem of water entry across the gills by osmosis. The additional problem of diffusional loss of salts is minor in comparison; cell membranes can be virtually impermeable to salts and passive loss of salts from FW fishes is very low. Gills must have a large surface area permeable to oxygen (and consequently also to water) with a short diffusion distance to the blood. Oxygen saturation and temperature can be very variable in FW (in contrast to the stable marine environment) and atmospheric oxygen content was probably lower when the vertebrate ancestors entered FW (Berner and Canfield, 1989), exacerbating the problems of oxygen uptake. Efficient gills (in the respiratory, not the osmoregulatory sense) would have been essential, belying Homer Smith's idea that the dermal armour of

Osmoregulation and Drinking in Vertebrates, edited by N. Hazon & G. Flik.
© 2002 BIOS Scientific Publishers Ltd, Oxford

the early fish would have protected them from the osmotic influx (Smith, 1953). In present day river lampreys, the urine flow rate needed to eliminate the excess water represents 20% of the body weight per day at 10°C, with a Q_{10} of five for acute temperature change (McVicar and Rankin, 1983).

In response to this challenge, lampreys and all the jawed vertebrates, but not hagfish, evolved lowered body fluid salt concentrations, which limited osmotic water entry from FW by reducing the plasma osmolality to between a quarter and a third that of SW. This is apparently the maximum reduction consistent with maintenance of intracellular potassium concentration (Burton, 1968). Even so, efficient filtration organs are required to produce urine at the high rates seen in present day FW fishes. Hagfish have a small number (around 70) of very large, segmentally-arranged glomeruli (Fänge, 1963) which seem to produce primary urine by secretion at physiological blood pressures (Riegel, 1998, 1999) but filter at elevated perfusion pressures (Riegel, 1978). Hagfish are confined to SW and cannot osmoregulate, although the Atlantic hagfish *Myxine glutinosa* can be gradually acclimated to salinities as low as 600 mOsm kg^{-1} and can be maintained in good condition for many weeks in 630 mOsm kg^{-1} (Cholette *et al.*, 1970).

Lampreys and gnathostomes need many more glomeruli to volume-regulate whilst osmoregulating in FW. A 60 g river lamprey has over 2000 glomeruli (Logan *et al.*, 1980b), all continually filtering (Brown and Rankin, 1999). Although there are a few aglomerular teleost species that produce urine by tubular secretion, the low number of glomeruli in hagfish is likely to be the primitive condition rather than a secondary reduction, in view of their segmental arrangement. Their kidneys are best seen as essential pre-adaptations to FW entry, increased filtering capacity having evolved gradually in brackish water estuaries. However, the recent discovery of hagfish fossils in what were thought to be FW Carboniferous deposits means that we cannot be certain that hagfish have always been confined to SW (Poplin *et al.*, 2001).

As osmoconformers, hagfish are virtually isoosmotic with their environment, although there is some regulation of salt concentrations. Urine flow is very low, but how is the excreted fluid replaced? A very slight plasma hyperosmocity would be sufficient for small volumes of water to enter across the gills, in view of their very high water permeability (Rudy and Wagner, 1970). Indeed, over the range 600–1540 mOsm kg^{-1}, *M. glutinosa* serum was slightly (0–2%) hyperosmotic (Cholette *et al.*, 1970). An alternative way to replace lost water would be by drinking.

The lampreys and gnathostomes that recolonized the marine environment, having evolved reduced body fluid osmolalities in FW, faced a new osmotic challenge. Two strategies evolved on more than one occasion; use of urea (and trimethylamine oxide) to raise body fluid osmolality to that of SW (or slightly higher) and hypo-osmotic regulation. The lampreys adopted the latter (as did the teleosts) whilst their nearest relatives, the elasmobranchs use the former. The energetic costs of the two strategies are similar (Kirschner, 1993), which probably explains the persistence of both.

Hypo-osmotic regulation involves replacing osmotic loss by drinking SW. This results in absorption of salts, partly by passive diffusion and partly to facilitate solute-linked water transport. The excess salts, together with any which enter by diffusion across the integument, are secreted across the gills. The ability to drink on re-entering the marine environment implies that drinking responses persisted during a long period of evolution in FW. It used to be thought that FW fish do not drink, although there have long been reports to the contrary (Maetz and Skadhauge, 1968).

The question of why fish should drink FW when large volumes of water are continually entering across their gills, will be discussed later in the chapter.

2. Drinking in hagfishes

Morris (1960) reported that under experimental conditions some Atlantic hagfish, such as *M. glutinosa* swallowed quite large amounts of SW, whilst others swallowed none at all. As there was no measurable absorption from the gut he concluded that the gut fluid and blood must be isotonic. Kobayashi *et al.* (1983) found that the saline-injected inshore hagfish, *Eptatretus burgeri,* drank 0.26 ± 0.06 ml kg^{-1} h^{-1} (n = 4). This rate was not increased by injection of angiotensin, although the groups given three different doses all drank (the amounts were: 0.36 ± 0.18, 0.44 ± 0.14 and 0.56 ± 0.23 ml kg^{-1} h^{-1}; n = 6, 7 and 7, respectively). Although these rates were very low they were not zero, which is what the same study found with the Arctic lamprey *Lampetra japonica japonica* in FW. Study of drinking vs. branchial water influx (if indeed both occur) in hagfish warrants further investigation.

3. Drinking in lampreys

The first demonstration of drinking in lampreys was in a group of 18 river lampreys (*Lampetra fluviatilis*) caught in FW early in their upstream spawning migration, and then acclimated to 50% SW (Morris, 1958). Drinking rate was calculated from accumulation of phenol red following blockage of the rectum. The four fish judged to have been osmoregulating successfully swallowed from 105 to 221 ml kg^{-1} day^{-1} and absorbed 76% of the imbibed fluid. This study was extended by Pickering and Morris (1970), who kept 14 river lampreys (*L. fluviatilis*) and three sea lampreys (*Petromyzon marinus*) in 33% SW for 2 days and, following ligature of the urinary papilla and blockage of the rectum, transferred them to 50% SW containing phenol red for 24 hours. Thirteen of the *Lampetra* drank from 2.8 to 99.5 ml kg^{-1} day^{-1}, of which 9.1–84.2% was absorbed. The *Petromyzon* drank from 9.2 to 53.5 ml kg^{-1} day^{-1} and absorbed from 58.7 to 80%. No mean values were given, perhaps not surprisingly in view of the variation encountered.

Everted and non-everted sacs from different regions of the intestines of FW river lampreys transferred to 50% SW for 24 hours were used to measure water flux (by weight change) and fluxes of ^{22}Na, ^{35}Cl, $^{35}SO_4$ and ^{45}Ca. Water was absorbed in the absence of an osmotic gradient in the anterior parts of the intestine, where there was a large active absorption of monovalent ions. Divalent ion fluxes were very low (Pickering and Morris, 1973). Rankin (1997) found that more than 99% of the sodium swallowed was absorbed by the intestines of river lampreys in full SW (one individual had less than 1 mmol l^{-1} in the rectum).

The difficulty with studying marine osmoregulation in migrating lampreys is that following FW entry, they progressively lose the ability to osmoregulate in even diluted SW. One reason is that the intestine degenerates, a process that is stimulated by gonadal hormones (Pickering, 1976; Larsen and Dufour, 1993) and can be partly reversed by hypophysectomy (Larsen, 1972). Pickering and Dockray (1972) found that gonadectomized river lampreys maintained the same drinking rate following transfer to 50% SW after being kept for 2 months in FW (125.2 ml kg^{-1} day^{-1} with

89% absorbed) as those starting their upstream migration (122.3 ml kg^{-1} day^{-1} with 95% absorbed). In contrast, sham-gonadectomized animals transferred to 50% SW after 2 months in FW drank very little (10.2 ml kg^{-1} day^{-1} with 64% absorbed). The net *in vitro* mucosal to serosal fluxes of sodium and chloride were high in gonadectomized lampreys after 24 hours immersion in 50% SW, but very much reduced in sham-operated animals (Pickering, 1978).

Kobayashi *et al.* (1983) failed to initiate drinking in FW-acclimated Japanese lampreys, *L. japonica*, by injection of 100 or 500 μg per kg body weight angiotensin II, which is dipsogenic in most vertebrates including teleost and elasmobranch fish. In an attempt to rule out the progressive degeneration of the intestine following FW entry as a reason for failure of angiotensin to induce drinking, Rankin *et al.* (2001) used both FW-acclimated and long-term brackish water-acclimated *L. fluviatilis*, 1 week after they had been transferred to FW. It was considered unlikely that significant degeneration had occurred in such a short time; at any rate drinking re-started following transfer back to brackish water after 1 week in FW. Several different angiotensins were injected, including homologous Asn^1Val^5-angiotensin II; but no stimulation of drinking was observed. However, the angiotensin converting enzyme inhibitor captopril (which inhibits formation of the active hormone, angiotensin II from angiotensin I in the blood) reduced drinking in 50% SW-acclimated lampreys, suggesting that further investigation was required. The smooth muscle relaxant papaverine, through its hypotensive action, stimulated the renin–angiotensin system to cause increased drinking in the European eel (Tierney *et al.*, 1995), but failed to stimulate drinking in these 50% SW-acclimated lampreys; in fact it caused a highly significant inhibition. The action of papaverine on lamprey blood pressure is unknown and captopril may also act independently of the renin–angiotensin system, as suggested in the Japanese eel (Takei and Tsuchida, 2000), so no evidence for the involvement of the renin–angiotensin system in the regulation of drinking in lampreys could be claimed.

As very few lampreys have been caught at sea, physiological studies have been largely confined to either adult upstream migrants (which are already losing their ability to osmoregulate in SW) or to the landlocked adults of the anadromous *P. marinus* which have their parasitic phase in FW in the Great Lakes. A few river lampreys are caught in coastal eel traps prior to their upstream migration. These osmoregulate well in SW and, in contrast to all other vertebrates studied apart from birds and mammals, are capable of producing hyperosmotic urine (Logan *et al.*, 1980a; Rankin, 1997; Brown and Rankin, 1999). Lampreys caught in the brackish water of the Ringkøbing Fjord in Denmark could be successfully re-acclimated to full-strength SW (1015 mOsm kg^{-1}) in which they drank at 7.39 ± 0.92 (standard error of mean) ml kg^{-1} h^{-1} ($n = 6$) at 12°C (= 177ml kg^{-1} day^{-1}) compared to 0.081 ± 0.036 ml kg^{-1} h^{-1} ($n = 6$) at 10°C in FW (Rankin, 1997).

In view of the paucity of information on lamprey drinking, further experiments were performed on lampreys in different salinities and following a change in salinity.

4. Materials and methods

4.1 *Measurement of drinking rate*

^{51}Cr-EDTA (Amersham) was used as an inert marker to estimate the volume of water swallowed. It has long been used for this purpose in teleosts and elasmobranchs

(Skadhauge and Lotan, 1974; Hazon *et al.*, 1989) but was validated for use in lampreys. River lampreys of 35–70 g were used. In view of this weight range, an attempt was made to find the weight exponent of drinking rate (which would be expected to be more related to gill surface area than to body mass). Log drinking rate (ml h^{-1} ΔOsm^{-1}; see below) was plotted against log body mass. An extremely poor correlation was obtained ($r^2 = 0.05$, $p = 0.36$, $n = 19$) so the exponent found ($mass^{0.46}$) was not used. Instead, attempts were made to ensure that the mean weight in each group was similar and drinking rates were expressed as millilitres per kilogram body weight per hour.

Five or six lampreys at a time were placed in a shallow bath containing 5–10 l of water from the lamprey stock tank, in which the bath was suspended to ensure a constant temperature of 10 or 12°C and allowed to settle down for approximately 1 h. Two to four MBq ^{51}Cr-EDTA solution was then added to the centre of the vigorously aerated tank without disturbing the lampreys. After 3–5 h immersion, the lampreys were quickly netted and transferred to a neighbouring rinse tank for 20–30 minutes, after which they were transferred individually to a 0.1 g l^{-1} MS222 (3-aminobenzoic acid ethyl ester methanesulphonate salt, Sandoz) anaesthetic bath. Lampreys can live for long periods in air and it proved possible to measure the total body radioactivity of anaesthetized animals coiled in the bottom of a 7 cm diameter counting vial using a Bicron Labtech gamma counter.

Counting the entire lamprey eliminated any possibility of loss of counts during ligation and excision of the gut, but it was essential to ensure that only radioactivity within the gut was counted. The rinse bath ensured removal of counts from the body surface, gill pouches, etc., and the few minutes in anaesthetic before respiratory movements ceased constituted a second rinse. To test if any marker had been absorbed into the lamprey bloodstream, large (>1 ml) blood samples taken from the caudal vessels of lampreys exposed to ^{51}Cr-EDTA for 24 h. These samples were counted in a Wallac 1480 Wizard gamma counter and gave background counts. Confirmation that contamination was successfully avoided came from some individuals in the FW group that also gave background counts. The FW group had a mean drinking rate of 0.0805 ml kg^{-1} h^{-1} with a standard deviation of 0.0815.

Not all water swallowed is absorbed; some serves to flush out non-absorbed divalent ions and is voided through the anus. If fish are left too long in ^{51}Cr-EDTA (or any marker) solution, counts will eventually be lost by this route; radioactivity eventually equilibrates which leads to an underestimation of drinking rate. SW-acclimated lampreys exposed to ^{51}Cr-EDTA for variable times between 2 and 5 h showed no tendency for calculated drinking rate to decrease with time of exposure, therefore a 3 h exposure time was employed. Drinking rates reported at individual time points are the mid-point of a single 3 hour immersion period in tanks containing ^{51}Cr labelled EDTA.

4.2 *Drinking rates in fish acclimated to different salinities*

Variation in drinking rate with salinity was studied in lampreys acclimated for more than 2 weeks to SW dilutions of the following osmolalities: 426, 522, 662 and 845 mOsm kg^{-1} at 10°C ($n = 6$ in each group). The osmotic gradient between plasma and bath water was calculated for each fish. Blood samples (ca. 0.5 ml) were taken from the caudal vessels of anaesthetized fish immediately after counting into 1 ml syringes rinsed out with 2 i.u. ammonium heparin (Leo Pharmaceutical Products, Denmark).

The blood was centrifuged for 4 min at 10 000*g* and the osmolality of 10 μl plasma samples was measured with a Wescor Vapro 5520 vapour pressure osmometer.

4.3 *Drinking rates following salinity transfer*

The time course of changes in drinking rates were studied following acute transfers from local (Great Belt) SW at 780 mOsm kg^{-1} to FW. Blood samples were taken immediately after the whole-body counting and plasma osmolality was measured. Direct transfer to full SW was not possible following acclimation to FW, so the following FW to brackish water transfers were attempted.

(i) Lampreys caught in Ringkøbing Fjord and held by the fisherman in 239 mOsm kg^{-1} water were acclimated to FW for 34 days. Seven fish were then transferred directly to 520 mOsm kg^{-1} brackish water in a 10 l tank. ^{51}Cr-EDTA was added to the water after 17 min. The lampreys were removed to a rinse tank 2 h 43 min later. Blood samples were not taken. In a separate group of these lampreys, five out of 15 were dead 1 day after the same transfer, so this experiment was discontinued.

(ii) Lampreys caught in Ringkøbing Fjord in 440 mOsm kg^{-1} water and held by the fisherman in 411 mOsm kg^{-1} water were acclimated to FW for 30 days then transferred to: (a) 457 mOsm kg^{-1} water for 6.25 h, (b) 458 mOsm kg^{-1} for 24 h, (c) 480 mOsm kg^{-1} water for 72 h, or (d) 450 mOsm kg^{-1} water for 72 h.

Blood samples were taken immediately after the whole-body counting and plasma osmolality was measured. Following examination of the resultant data, groups c and d were combined.

4.4 *Effects of hormones*

For treatment with hormones or saline, lampreys were netted, wrapped in a damp cloth, turned upside-down (which immobilized them as long as the head was kept covered), rapidly given a 0.5 ml intraperitoneal (i.p.) injection and released into the ^{51}Cr-EDTA bath. The entire procedure normally took only a few seconds. Use of anaesthesia was precluded as it inhibits drinking (Rankin, 1997). Experiments were carried out on both long-term FW-acclimated lampreys and on lampreys 1 week after transfer from brackish water to FW.

4.5 *Calculation of water absorption*

Absorption rates were measured in long-term acclimated lampreys, four in full SW (994 mOsm kg^{-1}) and four in 50% SW. Small samples of rectal fluid (ca. 0.25 ml) were taken with a polythene catheter attached to a syringe from fish under MS222 anaesthesia after exposure to ^{51}Cr-EDTA for 24 h. The samples were counted, as was a sample of the bath water, in a Wallac 1480 Wizard gamma counter and percentage water absorption calculated.

4.6 *Measurement of gill water permeabilities*

Diffusional water permeability was estimated in lampreys acclimated to SW (920 mOsm kg^{-1}, $n = 6$, body weight 37.6 ± 3.6 g) or FW ($n = 5$, body weight 30.6 ± 2.8 g).

They were injected (as described above) with 0.7 mBq (20 μCi) tritiated water (^3HHO, Amersham) and placed in a holding tank for 30 min to allow for absorption and distribution of the tracer. Each was then placed in a vigorously aerated, dark, 4 l bath fitted with polythene tubing, through which samples could be taken without disturbing the animal. Samples (1 ml) were taken at 30 min intervals for 5 h with a syringe attached to the tubing. This was then added to scintillation fluid and counted using a Packard Tricarb 2200CA liquid scintillation counter.

Water fluxes were calculated using a modification of the method of Motais *et al.* (1968). The turnover, λ, between body water and bath water (expressed as fraction exchanging per hour, not percentage per hour as originally stated) was estimated from a plot of log (counts at equilibrium – counts at time t) against time. Body water content was estimated by drying lampreys to constant weight in an oven at 80°C. Water outflux was calculated from:

$$\text{Outflux} = \lambda\,(V_{int} \times V_{ext}/V_{int} + V_{ext})$$

V_{int} and V_{ext} are the internal and external water volumes (in millilitres), respectively. It was not possible to calculate the diffusional permeability coefficient as the surface area of the gills (A) was not known, but in a comparison between diffusional and osmotic permeability coefficients, A cancels out. The internal water concentration was estimated to be 55.2 mmol cm^{-3}.

Osmotic water permeability was estimated from water absorption rate minus urine flow rate. Permeability was expressed in relation to the osmotic gradient across the gills; σ (the reflection coefficient) was assumed to be 1.0. The urine flow rates used in the calculations of osmotic/diffusional permeabilities were 5.76 ml kg^{-1} day^{-1} for the SW group (Rankin, 1997) and 200.7 ml kg^{-1} day^{-1} for the FW group (McVicar and Rankin, 1983). All data are presented as mean ± standard error of mean. For multiple comparisons, single factor analysis of variance (ANOVA) was used.

5. Results

5.1 *Long term acclimated lampreys*

The lampreys regulated well in all salinities from 426 to 845 mOsm kg^{-1} (*Fig. 1*), with no significant effect of water osmolality on plasma osmolality ($p = 0.22$, ANOVA).

Osmolalities from a different batch of FW acclimated lampreys are shown for comparison. In the higher salinities, plasma osmolalities were more variable (*see Fig. 1*). In view of this variation, the osmotic gradient for each individual was calculated and plotted against drinking rate (*Fig. 2*). Results from FW and SW-acclimated fish (Rankin, 1997) are included in the figure (but not in the regression) for comparison. Some outliers (solid circles in *Fig. 2*) were excluded from the correlations (see Discussion).

The solid line in *Fig. 2* shows the correlation between drinking rate and osmolality gradient, ΔOsm (water osmolality – plasma osmolality). The equation of the line was:

$$\text{Drinking rate} = 0.0127\Delta\text{Osm} - 0.89\ (r^2 = 0.81,\ p < 0.001)$$

The dotted line takes into account the volume of water absorbed, rather than the volume swallowed. The percentage absorbed was estimated to be 73.6 ± 3.6% (n = 4) for the SW-acclimated group and 75.9 ± 4.9% (n = 4) for the group acclimated to

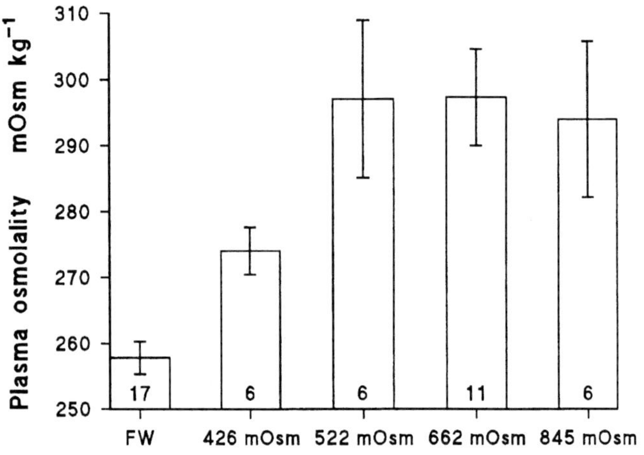

Figure 1. *Plasma osmolalities (means ± SEMs) of river lampreys acclimated to the four different dilutions of sea water used to determine the relationship between drinking rate and environmental osmolality shown in Figure 2 (water osmolalities given in mOsm kg^{-1}). The 622 mOsm kg^{-1} value is a mean from two groups that were acclimated to very similar water osmolalities. A freshwater group is included for comparison. Numbers in columns represent number of lampreys.*

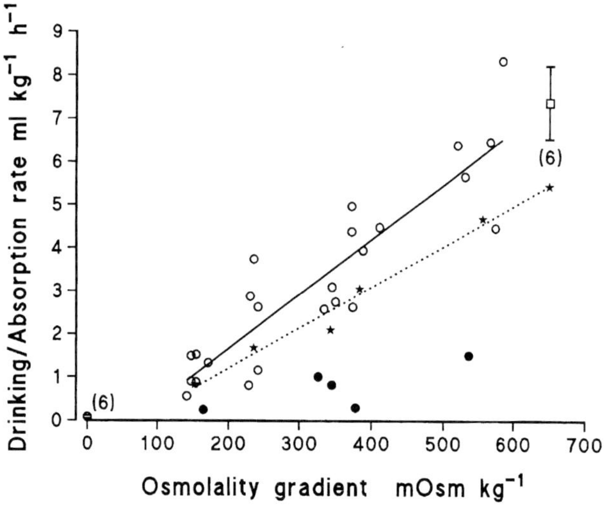

Figure 2. *Relationships betweeen osmotic gradient in mOsm kg^{-1} and drinking rate in ml kg^{-1} h^{-1} (open circles, solid line) and between osmotic gradient and calculated mean water absorption rate in ml kg^{-1} h^{-1} (stars, dotted line). Solid circles represent points excluded from the regressions for reasons given in the text. Values for a SW and a FW group are included for comparison (means ± SEMs, number of observations in parantheses). Values from the five experimental groups are included in the regression which produced the solid line; mean values for these five plus the SW group, all corrected for percentage absorption, are included in the regression which produced the dotted line.*

50% SW. As there was no significant difference, a mean value of 74.75% was used. Lampreys with similar osmotic gradients were grouped together and 74.75% of the mean drinking rates of each group (excluding the outliers mentioned above) was calculated. The equation of the linear regression line through these points was:

$$\text{Drinking rate} = 0.0094\Delta\text{Osm} - 0.69\ (r^2 = 0.98)$$

Results from a typical tritiated water flux experiment are illustrated in *Fig. 3*. Body water content and tritiated water flux data are shown in *Table 1*.

As the body water content was the same in FW- and 920 mOsm kg^{-1} SW-acclimated fish, the mean value (74.7 ± 0.87, $n = 6$) was used in all the calculations. The exchange rate of tritiated water was significantly faster ($p < 0.05$) and the outflux was significantly greater ($p < 0.05$) in the FW-acclimated lampreys. The osmotic water flux/diffusional water flux ratio was 0.92 in SW and 3.55 in FW.

Table 1 *Tritiated water flux data.*

Acclimation medium	Water content (% body mass)	$T_{1/2}$ (min)	Exchange rate (% h^{-1})	f_{out} (ml kg^{-1} min^{-1})	f_{out} (ml kg^{-1} min^{-1} Δmol^{-1})
920 mOsm	74.7 ± 0.8, $n = 3$	56.5 ± 4.0	54.6 ± 4.1	6.75 ± 0.50	0.122 ± 0.0091
Freshwater	74.7 ± 2.0, $n = 3$	44.2 ± 3.6*	69.7 ± 5.6*	8.73 ± 1.38*	0.156 ± 0.0127*

Data are means ± SEM, $n = 6$ for 920 mOsm and five for freshwater, unless stated otherwise. * = $p < 0.05$ (2-tailed t-test). $T_{1/2}$ = half time for exchange of body water; exchange rate = percentage of body water exchanging per hour; f_{out} = outflux of water (expressed without or with reference to the water concentration gradient across the gills, Δmol).

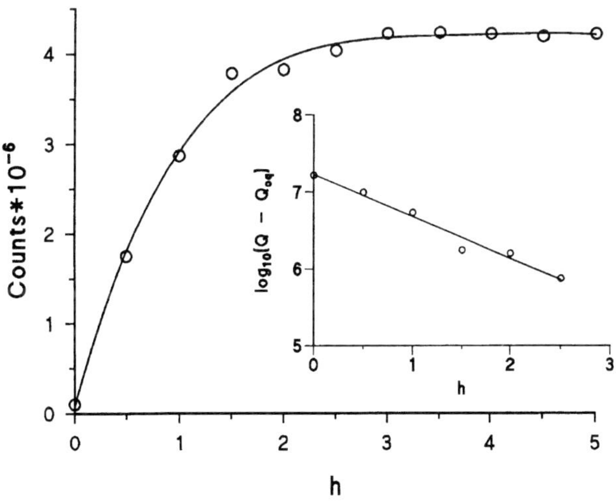

Figure 3. *Example of an experiment to determine tritiated water exchange in a freshwater-acclimated river lamprey. Total number of counts in the bath are plotted against time in hours and (insert) log_{10} (counts at time t – counts at equilibrium) is plotted against time in hours.*

5.2 *Acute transfer experiments*

Transfer from 780 mOsm kg^{-1} to FW produced a rapid fall in drinking rate (*Fig. 4*). Individual values are shown in the 2 h 9 min and 6 h 15 min post-transfer groups to show that some lampreys reduced drinking to very low levels over the period 15 minutes to 3 h following transfer. All were drinking at low rates by 48 h, with most of the change occurring within 24 h. Plasma osmolality (*see Fig. 4, insert*) appeared to be lower after 50 h but there was no significant effect of time on osmolality (ANOVA; $p = 0.32$).

Results of transfers from freshwater to brackish water (450–520 mOsm kg^{-1}) are plotted in *Fig. 5* (with groups c and d from series (ii) combined). As the bath osmolalities were similar but not identical and there was also variation in the plasma osmolalities, the osmotic gradients differed somewhat within the groups plotted. The figure shows that drinking rate rose quickly after transfer, the effect being complete within 24 h (*see Fig. 5*). Data from FW-acclimated lampreys are included for illustration but they are from a different batch of lampreys to those used in the transfer experiment. Plasma osmolality rose after transfer (*see Fig. 5, insert*), there being a significant change with time (ANOVA; $p < 0.05$, not including FW group).

5.3 *Hormone effects*

Results of a number of attempts to demonstrate stimulation of drinking in lampreys through administration of various angiotensins are assembled in *Fig. 6*. There was no significant effect on drinking rate from any of the treatments (ANOVA, $p = 0.47$).

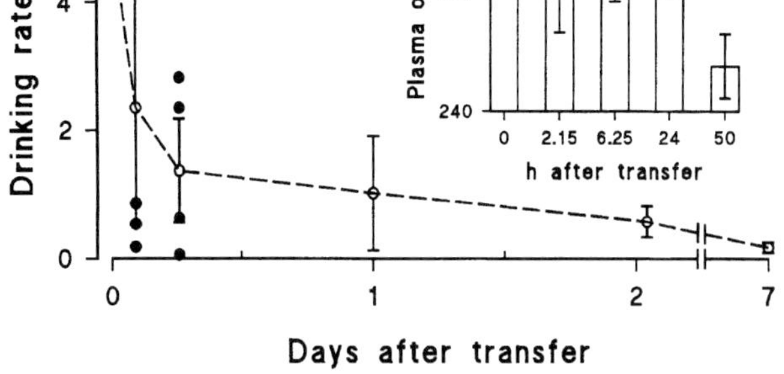

Figure 4. *Main figure: Drinking rate plotted against time (days) following transfer of river lampreys from 780 mOsm kg^{-1} brackish water to freshwater. Means (open circles) ± SEMs are shown. In addition, individual points for the 2h 9 min and 6h 15 min (mid-point times) groups are shown (closed circles). Data for drinking rate in one week freshwater-acclimated lampreys are included for comparison. Insert: Plasma osmolalities (number of observations given above columns, these also apply to the drinking rate data) in the experimental groups.*

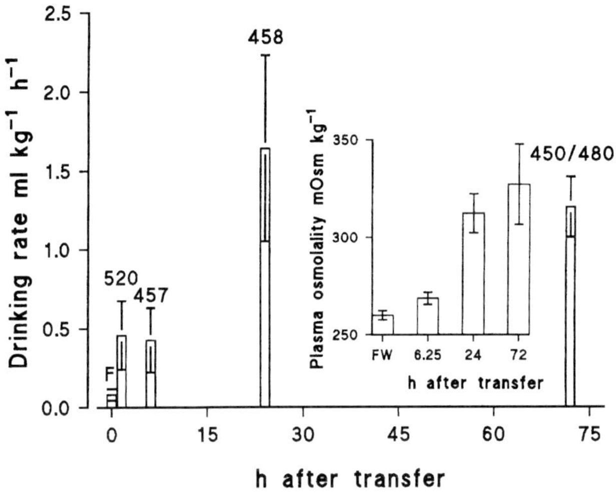

Figure 5. *Drinking rate plotted against time (hours) following transfer from FW to brackish water of the following osmolalities (given above columns in mOsm kg^{-1}): 520 (n = 7), 457 (n = 7), 458 (n = 7) and 450 or 480 (n = 4 of each). A freshwater control group (F; n = 15) is included for comparison. Plasma osmolalities are shown in the insert (same n as for drinking rates, osmolalities not measured in group transferred to 520 mOsm kg^{-1}).*

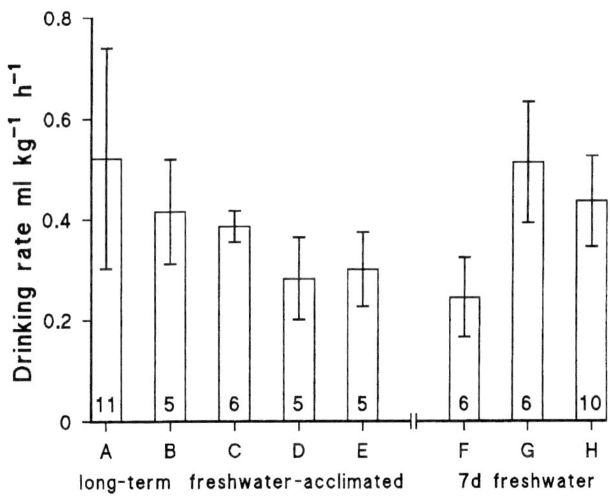

Figure 6. *Effects of angiotensin injections on drinking rate in FW-acclimated river lampreys. The left hand side of the figure shows effects of injection of the following: A. Control saline i.p.; B. Asp[1] Val[5] angiotensin II, 40 µg kg^{-1} i.p.; C. Asp[1] Val[5] angiotensin II, 169 µg kg^{-1} i.p.; D. Asn[1] Val[5] (lamprey) angiotensin II, 227 µg kg^{-1} i.v. and E. Asn[1] Pro[3] Ile[5] (dogfish) angiotensin II, 196 µg kg^{-1} i.p. into long-term FW-acclimated lampreys. The right hand side of the figure shows the following: F. non-injected control; G. saline-injected (i.p.) control and H. Asp[1] Val[5] angiotensin II, 168 µg kg^{-1}- injected (i.p.) lampreys following seven days acclimation to FW. Means ± SEMs are shown: numbers of observations are given in the columns.*

6. Discussion

Earlier work demonstrated that euryhaline lampreys (*L. fluviatilis* and *P. marinus*) osmoregulate in a very similar way to teleosts, by drinking SW whilst in hyper-osmotic environments. The difficulty of obtaining SW-acclimated lampreys meant that early studies were confined to lampreys in FW or after a very short time in 50% SW, so the discussion will centre on the results reported here from river lampreys that still retained their marine osmoregulatory capabilities.

The FW lampreys in this study drank at a very low rate (drinking rate 0.081 ± 0.036 ml kg^{-1} h^{-1}), which is in contrast to teleosts, many of which drink significant amounts of fluid in FW (Fuentes and Eddy, 1997). Why should they? Rainbow trout, *Oncorhynchus mykiss*, in very dilute (5 p.p.t.) brackish water drank when fed dry food, but drank little or nothing when fed moist food or when not fed (Ruohonen *et al.*, 1997). Even in SW, fed Atlantic salmon, *Salmo salar*, smolts drank significantly more than unfed fish (Usher *et al.*, 1988), suggesting a link between feeding and drinking, although the authors suggested a possible indirect relationship. Digestion in the gut requires fluid, and secretion into the alimentary canal by solute-linked water transport requires energy, some of which could be saved by drinking. In FW fishes it may cost almost as much to secrete water entering at the gills into the gut as to eliminate it by filtration, with reabsorption of the salts, renally. Water may be swallowed accidentally along with food, but drinking has often been recorded in non-feeding fish (Fuentes and Eddy, 1997). Hagfish, which do not osmoregulate, appear to drink but our knowledge of this is scanty. It is also of interest to note that marine elasmobranchs, which are iso-osmotic or slightly hyperosmotic to SW, can also drink (Hazon *et al.*, 1989; see Chapter 2). Adult river lampreys do not feed in FW, so would not need to drink for digestive purposes. The same applies in SW, where they imbibe liquid food. Lamprey drinking appears, therefore, to serve solely as an osmoregulatory function.

In the study of long-term acclimation to different salinities, a few lampreys had much lower counts in their guts than all other members of their groups. In most cases, these were the last fish to be netted from the rinse tank and transferred to the anaesthetic bath. When this was noted, samples of the anaesthetic bath were counted and found to contain sufficient radioactivity to affect the results. Water in the rinse bath was also radioactive but it was not, of course, possible to tell if the counts came from the external surfaces and the net used for the transfer or from the gut. It was concluded that lampreys may be able to void their gut contents when stressed (as each lamprey is removed from the rinse bath the others get more stressed). *Petromyzon* can swallow blood equivalent to 30% of their body weight per day (Farmer, 1980) and in river lampreys from Ringkøbing Fjord, sampled at indeterminate times (one or more days) after feeding, up to 16.8% of their body weight was accounted for by intestinal contents (Rankin, 1997). The existence of a gut-emptying mechanism to facilitate flight reactions may not be surprising. In subsequent experiments all fish were anaesthetized simultaneously (lampreys can be maintained under light anaesthesia for many hours).

An alternative possibility could be that lampreys do not drink continually and that some can go for several hours without drinking. Japanese eels fitted with an oesophageal cannula drink continually in SW (Takei *et al.*, 1998), but lampreys may be different. It is not immediately obvious how lampreys drink, since they appear to remain attached to the sides or bottom of the tank by their sucker. They breathe by tidal ventilation of the gill pouches, which are connected by a respiratory tube. This is isolated from the buccal cavity by the velum during attachment, to maintain suction. The oesophagus leads directly from the buccal cavity, by-passing the gill pouches (Hardisty, 1979).

River lampreys acclimated to SW and to 50% SW absorbed about 75% of the water swallowed, as did Japanese eels after 1 week's acclimation to SW (Oide and Utida, 1968). There was a linear relationship between absorption rate (with the very low values mentioned above excluded) and osmotic gradient. This is not as unsurprising as it may appear, in view of the results obtained by Isaia and Hirano (1975) on eel gills. They concluded that: 'the rate of water loss was found to vary with the gradient in a markedly non-linear fashion, and water flow in response to a given gradient decreased at higher osmolarities'. This tendency was especially marked in SW eel gills, and there was no difference in water loss between gills incubated in 60% SW and those incubated in 100% SW. However, they were working with isolated incubated gills and these may not reflect the *in vivo* situation (Rankin and Bolis, 1984). Maetz and Skadhauge (1968) found that eels acclimated to double-strength SW drank approximately twice as much as those in SW.

Drinking rates of lampreys in SW are high compared to most published values for teleosts. For example, the drinking rates measured in ten species by Perrott *et al.* (1992) were much lower (14–49% of the lamprey rate) with the exception of that of the sea scorpion, *Myoxocephalus scorpius*. However, sea scorpions studied in Kerteminde had much lower drinking rates (1.3 ± 0.3 ml kg^{-1} h^{-1} in 25 p.p.t. SW at 13°C; Breinbjerg and Rankin, unpublished observations).

High drinking rates reflect a high branchial water permeability (possibly high because tidal ventilation of the gills may be less efficient than the counter-current system of teleosts). Branchial water permeability can be calculated from tritiated water exchange, if the volumes of the internal and external compartments are known. Body water content did not differ between FW- and SW-acclimated lampreys, and the combined result of $74.7 \pm 0.87\%$ ($n = 6$) was very similar to the value of $75.6 \pm 0.51\%$ found for *P. marinus* (Thorson, 1959). Tritiated water exchange rate and efflux were significantly higher in FW- than in SW-acclimated fish, but the results must be interpreted with caution as the lampreys in the FW group were smaller and no weight exponent was estimated. Osmotic water fluxes are similar in absolute magnitude in FW-acclimated and SW-acclimated river lampreys, i.e. +198.8 ml kg^{-1} day^{-1} (McVicar and Rankin, 1983, corrected for FW drinking rate) and −171.6 ml kg^{-1} day^{-1} (calculated from Rankin, 1997), respectively. This is in spite of the osmotic gradient being 2.7 times greater across the SW than FW gills. Osmotic permeability of the gills is therefore clearly greater in FW-acclimated than in SW-acclimated lampreys. This is similar to the situation in teleosts (Isaia, 1984).

Tritiated water efflux can be used to calculate diffusional water permeability, when the surface area over which exchange occurs is known, but there are many sources of error, as is also the case with calculation of osmotic water permeability (Rankin and Davenport, 1981). Osmotic and diffusional flux measurements were made on lampreys from different batches and of different body weight. Actual permeabilities could not be calculated, as gill surface area was not known, but the ratio of diffusional to osmotic permeability was found to be approximately one (0.92) in SW-acclimated lampreys. In six species of teleosts in SW the ratio was 0.86

(excluding the very high value of seven for rainbow trout), again very close to one considering all the errors involved (Isaia, 1984). The picture is very different for FW-acclimated fish. The five FW teleosts listed by Isaia (1984) have a mean ratio of 2.72 and Motais *et al.* (1968) found values of 6.12 for *Carassius* and 3.16 for *Anguilla* in FW. The ratio for FW lampreys was 3.55. The simplest explanation for a high ratio (apart from the presence of unstirred layers, which gills have evolved to minimize) is that osmotic flow is not proceeding by diffusion but by bulk flow through aqueous channels (Rankin and Davenport, 1981). A possible conclusion is that aqueous channels are present in the gills of FW- but not SW-acclimated fish. However, Isaia (1984) cautions about applying simplistic considerations to a heterogeneous barrier such as gills, where osmotic and diffusional fluxes may occur through different cells and where high salt permeability through leaky junctions may invalidate the assumption that the reflection coefficient is unity.

Transfer to FW caused a rapid fall in drinking rate without a change in plasma osmolality. There was considerable variability in both drinking rate and plasma osmolality at 2.15 and 6.25 h, but there was no correlation between the two. Further work is needed to determine the stimulus for cessation of drinking, but an obvious candidate is change in blood volume. This may also be involved in the initiation of drinking following transfer to 50% SW where, although plasma osmolality increased, there was no correlation between drinking rate and plasma osmolality 6.25 h after transfer.

The results of both transfers show that lampreys can regulate their drinking in response to changes in environmental osmolality, but we know nothing about the mechanisms involved. In teleost fish the renin–angiotensin system plays a central role (see Chapters 2 and 5). Attempts to stimulate drinking by angiotensin injection in *L. japonica* and *L. fluviatilis* failed (Kobayashi *et al.*, 1983; Rankin *et al.*, 2001). Apart from problems of rapid intestinal degeneration following FW entry, it is possible that i.p. hormone injection did not result in a long enough elevation of plasma levels in relation to the 4 h drinking period. However, action over a shorter time would still increase the cumulative volume swallowed. Several teleost species respond to i.p. or even intraarterial bolus injection of homologous or heterologous angiotensin II (see list in Fuentes and Eddy, 1997). Some of the studies listed used intramuscular injection, but the advantages of this over i.p. injection were doubted in lampreys (Rankin *et al.*, 2001). Constant intravenous (i.v.) infusion, as carried out in flounders (Balment and Carrick, 1985) and Japanese eels (Tsuchida and Takei, 1999) would be preferable but is technically difficult in lampreys. Further attempts should be made, including use of lamprey angiotensin I (Rankin, Watanabe, Nakajima and Takei, unpublished), which should have a more prolonged action, before ruling out a dipsogenic action of angiotensin in lampreys.

Acknowledgements

The work of the author was supported by the Danish Natural Science Research Council. I thank Lone Morgen for technical assistance and Peter Langholm for help with the tritiated water flux experiments.

Work in progress (Brown, Cobb, Frankling and Rankin, unpublished observations) suggests that angiotensin II has a very short half life in lamprey blood. This fact may have contributed to the failure to observe effects of a single injection on drinking rate.

References

Balment, R.J. and Carrick, S. (1985) Endogenous renin–angiotensin system and drinking behavior in flounder. *Am. J. Physiol.* **248:** R157–R160.

Berner, R.A. and Canfield, D.E. (1989) A new model for atmospheric oxygen over phanerozoic time. *Am. J. Sci.* **289:** 333–361.

Brown, J.A. and Rankin, J.C. (1999) Lack of glomerular intermittency in the river lamprey *Lampetra fluviatilis* acclimated to sea water and following acute transfer to iso-osmotic brackish water. *J. Exp. Biol.* **202:** 939–946.

Burton, R.F. (1968) Cell potassium and the significance of osmolarity in vertebrates. *Comp. Biochem. Physiol.* **27:** 763–773.

Cholette, C., Gagnon, A. and Germain, P. (1970) Isosmotic adaptation in *Myxine glutinosa* L.-1. Variations of some parameters and role of the amino acid pool of the muscle cells. *Comp. Biochem. Physiol.* **33:** 333–346.

Farmer, G.J. (1980) Biology and physiology of feeding in adult lampreys. *Can. J. Fish. Aquat. Sci.* **37:** 1751–1761.

Fänge, R (1963) Structure and function of the excretory organs. In: *The Biology of Myxine* (eds A. Brodal and R. Fänge). Universitetsforlaget, Oslo, pp. 516–529.

Fuentes, J. and Eddy, F.B. (1997) Drinking in marine, euryhaline and freshwater teleost fish. In: *Ionic regulation in animals: a tribute to Professor W.T.W. Potts* (eds N. Hazon, F.B. Eddy and G. Flik). Springer Verlag, Berlin, pp. 135–149.

Hardisty M.W. (1979) *Biology of the cyclostomes.* Chapman and Hall, London.

Hazon, N., Balment, R.J., Perrott, M. and O'Toole, L.B. (1989) The renin–angiotensin system and vascular dipsogenic regulation in elasmobranchs. *Gen. Comp. Endocrinol.* **74:** 230–235.

Isaia, J. (1984) Water and nonelectrolyte permeation. In: *Fish Physiology vol. X, Gills. Part B, Ion and water transfer* (eds W.S. Hoar and D.J. Randall). Academic Press, Orlando, Fla, pp. 1–38.

Isaia, J. and Hirano, T. (1975) Effect of environmental salinity change on osmotic permeability of the isolated gill of the eel, *Anguilla anguilla* L. *J. Physiol., Paris* **70:** 737–747.

Janvier, P. (1999) Catching the first fish. *Nature* **402:** 21–22.

Kirschner, L.B. (1993) The energetics of osmotic regulation in ureotelic and hypoosmotic fishes. *J. Exp. Zool.* **267:** 19–26.

Kobayashi, H., Uemura, H., Takei, Y., Itatsu, N., Ozawa, M. and Ichinohe, K. (1983) Drinking induced by angiotensin II in fishes. *Gen. Comp. Endocrinol.* **49:** 295–306.

Larsen, L.O. (1972) Endocrine control of intestinal atrophy in normal lampreys and of intestinal hypertrophy in gonadectomized lampreys, *Lampetra fluviatilis. Gen. Comp. Endocrinol.* **18:** 602.

Larsen, L.O. and Dufour, S. (1993) Growth, reproduction and death in lampreys and eels. In: *Fish Ecophysiology* (eds J.C. Rankin and F.B. Jensen). Chapman and Hall, London, pp. 72–104.

Logan, A.G., Moriarty, R.J. and Rankin, J.C. (1980a) A micropuncture study of kidney function in the river lamprey, *Lampetra fluviatilis*, adapted to sea water. *J. Exp. Biol.* **85:** 239–247.

Logan, A.G., Moriarty, R.J., Morris, R. and Rankin, J.C. (1980b) The anatomy and blood system of the kidney in the river lamprey, *Lampetra fluviatilis. Anat. Embryol.* **158:** 245–252.

McVicar, A.J. and Rankin, J.C. (1983) Renal function in unanaesthetized river lampreys (*Lampetra fluviatilis* L.): effects of anaesthesia, temperature and environmental salinity. *J. Exp. Biol.* **105:** 351–362.

Maetz, J. and Skadhauge, E. (1968). Drinking rates and gill ionic turnover in relation to external salinities in the eel. *Nature* **217**, 371–372.

Martini, F.H. (1998) The ecology of hagfishes. In: *The biology of hagfishes* (eds J.M. Jørgensen, J.P. Lomholt, R.E. Weber and H. Malte). Chapman and Hall, London, pp. 57–77.

Morris, R. (1958) The mechanism of marine osmoregulation in the lampren (*Lampetra fluviatilis* L.) and the causes of its breakdown during the spawning migration. *J. Exp. Biol.* **35:** 649–665.

Morris, R. (1960) General problems of osmoregulation with special reference to cyclostomes. *Symp. Zoo. Soc. Lond.* **1:** 1–16.

Motais, R., Isaia, J., Rankin, J.C. and Maetz, J. (1968) Adaptive changes of the water permeability of the teleostean gill epithelium in relation to external salinity. *J. Exp. Biol.* **51:** 529–546.

Oide, H. and Utida, S. (1968) Changes in intestinal absorption and renal excretion of water during adaptation to sea-water in the Japanese eel. *Mar. Biol.* **1:** 172–177.

Perrott, M.N., Grierson, G.E., Hazon, N. and Balment, J. (1992) Drinking behaviour in seawater and freshwater teleosts, the role of the renin–angiotensin system. *Fish Physiol. Biochem.* **10:** 161–168.

Pickering, A.D. (1976) Stimulation of intestinal degeneration by oestradiol and testosterone implantation in the migrating river lamprey, *Lampetra fluviatilis* L. *Gen. Comp. Endocrinol.* **30:** 340–346.

Pickering, A.D. (1978) Physiological aspects of the life cycle of the river lamprey, *Lampetra fluviatilis* L. *Annu. Rep. Freshwater Biological Association* **46:** 40–46.

Pickering, A.D. and Dockray, G.J. (1972) The effects of gonadectomy on osmoregulation in the migrating river lamprey: *Lampetra fluviatilis. Comp. Biochem. Physiol.* **41A:** 139–147.

Pickering, A.D. and Morris, R. (1970) Osmoregulation of *Lampetra fluviatilis* L. and *Petromyzon marinus* (Cyclostomata) in hyperosmotic solutions. *J. Exp. Biol* **58:** 165–175.

Pickering, A.D. and Morris, R (1973) Localisation of ion transport in the intestine of the migrating river lamprey *Lampetra fluviatilis* L. *J. Exp. Biol.* **58:** 165–176.

Poplin, C., Sotty, D. and Janvier, P. (2001) Un myxinoïde (Craniata, Hyperotreti) dans le Konservat-Lagerstätte Carbonifère supérieur de Montceau-les-Mines (Allier, France). *Compt. rend. Acad. Sci. ser.* IIA **332:** 345–350.

Rankin, J.C. (1997) Osmotic and ionic regulation in cyclostomes. In: *Ionic regulation in animals: a tribute to Professor W.T.W. Potts* (eds N. Hazon, F.B. Eddy and G. Flik). Springer Verlag, Berlin, pp. 50–69.

Rankin, J.C. and Bolis, L. (1984) Hormonal control of water movement across the gills. In: *Fish Physiology vol. X, Gills. Part B, Ion and water transfer* (eds W.S. Hoar and D.J. Randall). Academic Press, Orlando, Fla, pp. 177–201.

Rankin, J.C. and Davenport, A. (1981) *Animal Osmoregulation.* Blackie, Glasgow.

Rankin, J.C., Cobb, C.S., Frankling, S.C. and Brown, J.A. (2001) Circulating angiotensins in the river lamprey, *Lampetra fluviatilis*, acclimated to freshwater and seawater: possible involvement in the regulation of drinking. *Comp. Biochem. Physiol.* **129B:** 311–318.

Riegel, J.A. (1978) Factors affecting glomerular function in the Pacific hagfish *Eptatretus stouti* (Lockington). *J. Exp. Biol.* **732**, 261–267.

Riegel, J.A. (1998) An analysis of the function of the glomeruli of the hagfish mesonephric kidney. In: *The biology of hagfishes* (eds J.M. Jørgensen, J.P. Lomholt, R.E. Weber and H. Malte). Chapman and Hall, London, pp. 364–377.

Riegel, J.A. (1999) Secretion of primary urine by glomeruli of the hagfish kidney. *J. Exp. Biol.* **202:** 947–955.

Rudy, P.P. and Wagner, R.C. (1970) Water permeability in the Pacific hagfish *Polistotrema stoutii* and in the staghorn sculpin *Leptocottus armatus. Comp. Biochem. Physiol.* **34:** 399–403.

Ruohonen, K., Grove, D.J. and McIlroy, J.T. (1997) The amount of food ingested in a single meal by rainbow trout offered chopped herring, dry and wet diets. *J. Fish Biol.* **51:** 93–105.

Shu, D-G., Luo, H-L., Conway Morris, S., *et al.* (1999) Lower Cambrian vertebrates from south China. *Nature* **402:** 42–46.

Skadhauge, E. and Lotan, R. (1974) Drinking rate and oxygen consumption in the euryhaline teleost *Aphanius dispar* in waters of high salinity. *J. Exp. Biol.* **60:** 547–556.

Smith, H.W. (1953) *From fish to philosopher.* Little, Brown and Co., Boston, Mass.

Takei, Y. and Tsuchida, T. (2000) Role of the renin–angiotensin system in drinking of seawater-adapted eels *Anguilla japonica*: a reevaluation. *Am. J. Physiol.* **279:** R1105–1111.

Takei, Y., Tsuchida, T. and Tanakadete, A. (1998) Evaluation of water intake in seawater adaptation in eels using a synchronized drop counter and pulse injector system. *Zool. Sci.* **15:** 677–682.

Thorson, T.B. (1959) The partitioning of body water in sea lamprey. *Science* **130:** 99–100.

Tsuchida, T. and Takei, Y. (1999) A potent dipsogenic action of homologous angiotensin II infused at physiological doses in eels. *Zool. Sci.* **16:** 479–483.

Tierney, M.L., Cramb, G. and Hazon, N. (1995) Stimulation of the renin–angiotensin system, and drinking by papaverine in the seawater eel, *Anguilla anguilla*. *J.Fish Biol.* **46:** 721–724.

Usher, M.L., Talbot, C. and Eddy, F.B. (1988) Drinking in Atlantic salmon smolts transferred to seawater and the relationship between drinking and feeding. *Aquaculture* **73:** 237–246.

2

The control of drinking in elasmobranch fish with special reference to the renin–angiotensin system

W.G. Anderson, A. Wells, Y. Takei and N. Hazon

1. Introduction

Sea-water (SW) acclimated elasmobranchs adopt an unusual osmoregulatory strategy in that body fluids are maintained iso- or slightly hyper-osmotic to that of the surrounding medium (*Figure 1*). Plasma osmolality is elevated through increased levels of sodium and chloride ions, with respect to SW teleost fish, but also through the retention of high concentrations of urea (Smith, 1936). In higher vertebrates urea is considered a waste product following protein catabolism and is excreted. Elasmobranchs, however, actively synthesize and retain urea along with methylamines such as trimethylamine oxide (TMAO) and maintain an optimal intracellular ratio of 2:1 (urea:methylamines) (Yancey and Somero, 1980), to combat the toxic effects of urea. Investigation into the regulation of these osmolytes in SW elasmobranchs has focused on three tissues, the gills, the rectal gland (a specialized salt secreting organ) and the kidney.

2. Regulation of salt and water balance

2.1 *Gill*

There is a paucity of information regarding the transport of urea across elasmobranch gill epithelium but despite what has been termed as a 'diffusional barrier to urea loss' (Boylan, 1967), the gills remain the major site for passive urea efflux from the body fluids to the environmental medium. Mitochondrial rich chloride cells have been identified on the gill epithelium (Doyle and Gorecki, 1961), however, Na^+,K^+,ATPase activity was found to be 10–15 times less than that observed in teleost fish (Jampol and Epstein, 1970). As a specific role for chloride cells in the elasmobranch gill

Osmoregulation and Drinking in Vertebrates, edited by N. Hazon & G. Flik.
© 2002 BIOS Scientific Publishers Ltd, Oxford

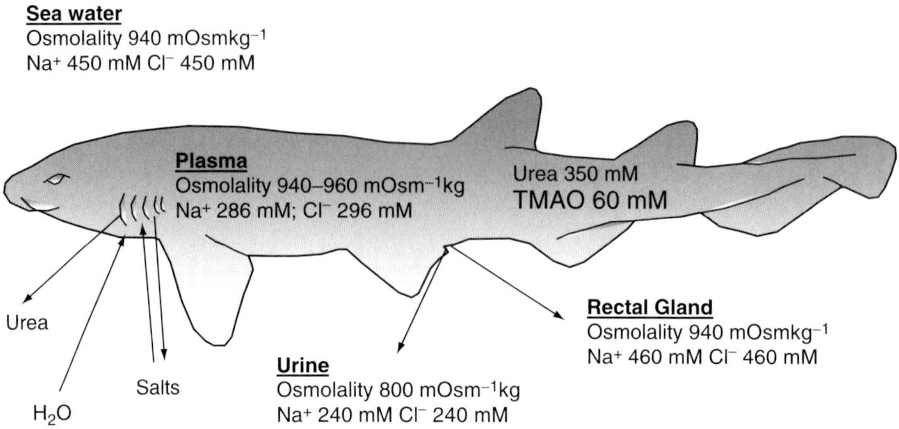

Figure 1. *Summary of osmolyte regulation in* Scyliorhinus canicula *acclimated to 100% sea water (SW).*

epithelium has yet to be identified (Laurent and Dunel, 1980) it appears that the gills do not play a major regulatory role with respect to plasma sodium and chloride balance in SW elasmobranchs. Finally, the iso/hyper-osmotic strategy adopted by SW elasmobranch fish generates a small passive influx of water. Indeed it was this movement of water into the animal that led researchers to believe that SW elasmobranchs did not have the physiological requirement to drink (Schmidt-Nielsen, 1997).

2.2 *Rectal gland*

Elasmobranchs are equipped with a specialized salt secreting organ, the rectal gland, which is supplied by a single artery and empties via a single duct into the posterior end of the intestine, close to the rectum. The gland secretes a fluid that is iso-osmotic to plasma but is almost entirely composed of sodium and chloride (Burger and Hess, 1960). Rectal gland activity is known to be intermittent in nature (Burger, 1962; Kent and Olson, 1982) and secretion rate has been shown to be closely related to volume expansion (Solomon *et al.*, 1985). Activity of the rectal gland could therefore be primarily associated with volume/salt loading during acclimation to varying environmental salinities or following single feeding events (Hazon *et al.*, 1997).

2.3 *Kidney*

SW elasmobranchs do not appear to be capable of producing a hyper-osmotic urine with respect to blood plasma. However, the elasmobranch nephron has been shown to be very complex with many functional regions and different countercurrent loops (Hentschel, 1987; Lacy and Reale, 1995). The principle role of the elasmobranch nephron appears to be the retention of urea through both passive and active routes (Boylan, 1972; Lacy and Reale, 1995; Lacy *et al.*, 1985; Smith and Wright, 1999). As a consequence the kidney may not therefore be a major site for regulation of plasma sodium and chloride ions. Identification of a juxtaglomerular (JG) apparatus in the kidneys of elasmobranchs (Lacy and Reale, 1990) has provided evidence for the presence of a renin–angiotensin sytem (RAS). In higher vertebrates a major physiological function of the RAS is the control of drinking.

2.4 *Intestine*

Given that for many years it was thought that there was no necessity for SW elasmobranchs to drink, the role of the intestine in osmoregulation has received little attention. However, recent research has demonstrated that SW elasmobranchs will drink the environmental medium under the appropriate osmotic conditions (Anderson *et al.*, 2002; Hazon *et al.*, 1997). As a consequence the intestine of elasmobranchs may well play a crucial and integrated role with renal and extra-renal tissues in the regulation of plasma salt and water balance.

3. The renin–angiotensin system – comparative aspects

The RAS is a peptide cascade system found in all classes of vertebrates investigated to date (*Figure 2*). Angiotensinogen is cleaved by renin to produce the decapeptide, angiotensin I (ANG I). The last two amino acids of the sequence are then removed by angiotensin converting enzyme (ACE) to produce the octapeptide angiotensin II (ANG II) which is considered to be the principle bioactive component of the cascade. ANG II is further broken down by amino peptidases to produce angiotensins III and IV. Evidence is now emerging that both these products exert some degree of bioactivity in the vertebrate classes investigated to date.

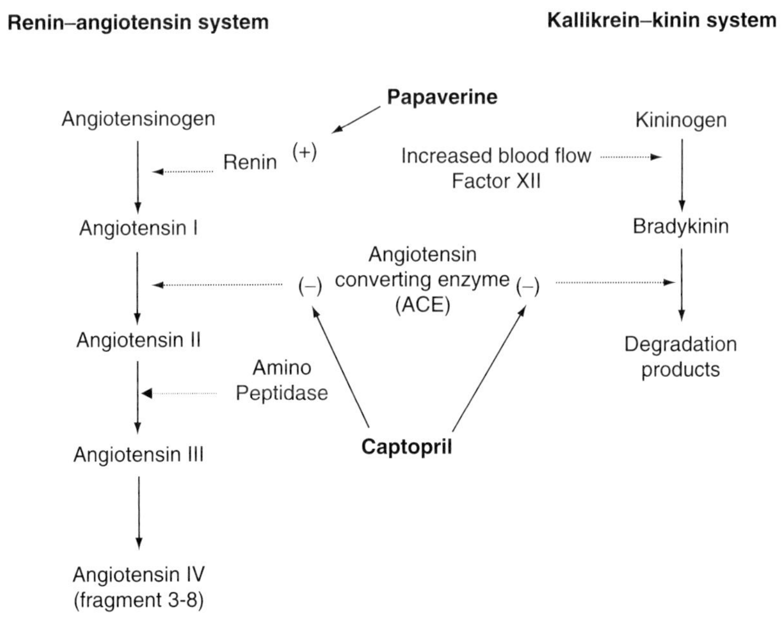

Figure 2. *Illustration of the renin–angiotensin system and the kallikrein–kinin system showing the actions of the angiotensin converting enzyme inhibitor, captopril and the smooth muscle relaxant, papaverine.*

There is a high degree of structural homology in ANG I throughout the vertebrate phyla with only amino acids at position 1, 5 and 9 showing heterology (*Table 1*). Aspartic acid and asparagine interchange at position 1 and valine and isoleucine at position 5. However, in elasmobranchs there is an additional unique substitution of proline for valine at position 3 (Takei *et al.*, 1993). It is this substitution that may explain why, for many years, the use of heterologous bioassays failed to demonstrate evidence for the presence of a functional RAS in elasmobranchs (Bean, 1942; Nishimura, 1985; Nishimura *et al.*, 1970).

In mammals, ANG II is known to be a potent vasopressor and dipsogenic peptide and is indirectly sodium-reabsorptive through a steroidogenic action on the adrenal gland increasing secretion of aldosterone. Description of a RAS in elasmobranchs and the physiological functions of ANG II has only recently been possible with the purification and identification of ANG I from the Japanese dogfish, *Triakis scyllia* (Takei *et al.*, 1993). The vasopressor and steroidogenic actions of the RAS in elasmobranchs have recently been reviewed (Anderson *et al.*, 2001a; Hazon *et al.*, 1999) so this chapter will concentrate primarily on the dipsogenic actions of the RAS in elasmobranch fish.

4. Control of drinking in elasmobranchs

Previously, intra-vascular and intra-peritoneal administration of heterologous ANG II did not induce a drinking response in elasmobranch fish (Hazon *et al.*, 1989; Kobayashi *et al.*, 1983). However, administration of the smooth muscle relaxant, papaverine, in the European lesser spotted dogfish, *Scyliorhinus canicula*, significantly increased drinking rate in these fish. Furthermore, the ACE inhibitor captopril when co-administered with papaverine reduced the papaverine-induced response (Hazon *et al.*, 1989). The hypotensive effect of papaverine is thought to induce a

Table 1. *Amino acid sequences of angiotensin I from a variety of vertebrate classes*

	Amino acid sequence									
Species	1	2	3	4	5	6	7	8	9	10
Human, pig, rat, rabbit, horse, dog	Asp	Arg	Val	Tyr	Ile	His	Pro	Phe	His	Leu
Bovine	Asp				Val				His	
Chicken, quail	Asp				Val				Ser	
Turtle	Asp				Val				His	
Alligator	Asp				Val				Ala	
Snake	Asx				Val				Ser	
Bullfrog	Asp				Val				Asn	
Lungfish	Asn				Val				Thr	
Goosefish	Asn				Val				His	
Flounder	Asn				Ile				Asn	
Salmon	Asn				Val				Asn	
Eel	Asn				Val				Gly	
Bowfin	Asp				Val				Asn	
Dogfish	Asn		Pro		Ile				Gln	

Sources: Kobayashi and Takei (1996); Joss *et al.* (1999); Takei *et al.* (1998).

release of endogenous renin (Nishimura *et al.*, 1979) which leads to the activation of the RAS cascade. Co-administration of captopril acts to block ACE and hence the conversion of ANG I to ANG II and, therefore, inhibits the papaverine-induced activation of the RAS. As a consequence, Hazon *et al.* (1989) postulated that the endogenous RAS in *S. canicula* was involved in the control of drinking. Recently, a dose-dependent increase in drinking to intra-vascularly administered elasmobranch ANG II was observed in both *S. canicula* and *T. scyllia* (Anderson *et al.*, 2001c). It is clear, therefore, that the RAS is closely involved in the control of drinking in elasmobranchs and that ANG II acts as a potent dipsogenic agent as with higher vertebrates. It is also worth noting that an additional action of ACE is in the control of bradykinin (BK) degradation (Gardiner *et al.*, 1993) (*Figure 2*). BK is an active component of the kallikrein–kinin system and has been shown to be potently antidipsogenic in the eel (Takei *et al.*, 2001). Therefore, the observed decrease in drinking following captopril injection may be due to an inhibition of ANG II production in combination with an increase in circulating levels of BK; however to date, elasmobranch BK has not been isolated or characterized (Conlon, 1999).

5. Control of drinking and osmotic balance

5.1 *Comparative aspects*

Transfer of the Japanese eel, *Anguilla japonica*, from the hypo-osmotic environment of fresh water (FW) to the hyper-osmotic environment of SW caused an immediate but transient increase in drinking rate (Hirano, 1974). This increase, termed the 'chloride response' was shown to be triggered by the presence of chloride ions in SW which was detected in the buccal cavity of eels (Hirano, 1974). Following the initial peak in drinking there is a sustained increase in drinking, approximately 10 times greater than that observed in FW eels (Tierney *et al.*, 1995). Concomitant with this sustained increase in drinking in SW eels is an increase in circulating levels of ANG II (Okawara *et al.*, 1987; Takei *et al.*, 1988b). This increase in plasma ANG II is believed to be the primary regulating factor in the increased drinking rate of SW-acclimating euryhaline eels (Tierney *et al.*, 1995; Tsuchida and Takei, 1999). In mammals and birds the order of potency for dipsogenesis is cellular dehydration followed by ANG II followed by extra-cellular dehydration (Fitzsimons, 1979; Takei, 2000). Perhaps not surprisingly, given their close association with an aqueous environment, aquatic vertebrates respond to these stimuli in a different manner.

Cellular dehydration in animals can be induced through the intra-vascular administration of hyper-osmotic fluids such as NaCl or large sugar molecules such as sucrose or mannitol. NaCl and sucrose occupy the extra-cellular fluid compartment and draw water from the cellular compartment, causing a profound cellular dehydration and an associated extra-cellular hypervolaemia. However, although mannitol induces a similar fluid shift, gradual equilibration of mannitol between the two compartments does occur (Fitzsimons, 1979). Injection of hypertonic NaCl into FW *A. japonica*, causing a theoretical cellular dehydration between 2.8 and 5.7%, consistently inhibited drinking (Takei *et al.*, 1988b). However, controlled haemorrhage in *A. japonica* inducing extra-cellular dehydration caused a significant increase in drinking (Hirano, 1974) and increase in plasma ANG II levels (Kobayashi and Takei, 1996). Furthermore, this induced increase in drinking could be completely blocked by the ACE inhibitor, captopril (Kobayashi and Takei, 1996). It is clear, therefore, that unlike mammals and birds, extra-cellular dehydration is the major stimulus for drinking in eels, followed by an increase in plasma levels of ANG II, with cellular dehydration, if anything, demonstrating an anti-dipsogenic effect (Takei *et al.*, 1988b; Takei, 2000).

5.2 *Environmental transfer of elasmobranchs*

Recently the mechanisms behind the drinking response in elasmobranchs were investigated by Anderson *et al.* (2002). Two species of elasmobranch, *T. scyllia* and *S. canicula*, were acclimated to the reduced environmental salinity of 80% then rapidly transferred to 100% SW. As the fish were transferred to the hyper-osmotic environment of 100% SW plasma sodium and chloride levels increased significantly within 6 hours, however, plasma urea remained unchanged. Therefore, the observed increase in plasma osmolality was not solely due to osmotic water loss. Drinking rate was measured in both species using an oesophageal cannulation technique (Anderson *et al.*, 2001c; Takei *et al.*, 1998), and was found to be maximal and significantly greater within 2–3 hours post transfer. Therefore, the increase in plasma osmolality was attributable in part to osmotic water loss but also to an increase in drinking rate.

Plasma ANG II levels were assessed using a similar method to those previously described (Bernier *et al.*, 1999; Tierney *et al.*, 1998). However, it is worth noting that the antibody used in the radioimmunoassay (Yamaguchi, 1981) demonstrated equal cross reactivity for ANG II and ANG III. Therefore, it must be recognized that reported values for ANG II may also include ANG III. The increase in drinking observed following transfer from 80 to 100% SW (Anderson *et al.*, 2002) was mirrored by a significant increase in plasma ANG II levels (*Figure 3*). ANG II levels increased rapidly within 90 minutes of transfer and returned to basal levels within 3 hours. An emergency role for ANG II in the control of blood pressure following acute volume loss and/or reduction in blood pressure has previously been suggested (Anderson *et al.*, 2001a; Bernier *et al.*, 1999; Kobayashi and Takei, 1996). Such a role for ANG II in the control of drinking during environmental salinity acclimation in elasmobranchs may also be possible given the rapid but transient increase in plasma levels of ANG II (*Figure 3*).

5.3 *Extra-cellular dehydration in elasmobranchs*

To determine the stimulus for the observed drinking response following environmental acclimation Anderson *et al.* (2002) induced cellular and extra-cellular dehydration in *S. canicula* acclimated to 100% SW. Extra-cellular dehydration in *S. canicula* was induced through controlled haemorrhage of approximately 5.7% total blood volume. Following haemorrhage there was a rapid but transient increase in drinking rate of nearly 40-fold over basal levels. Plasma ANG II levels also increased above basal levels taken just prior to haemorrhaging. However, although the increase was not statistically significant, levels remained elevated for up to 3 hours post haemorrhage (*Figure 4*). It is apparent therefore that the increase in drinking following transfer from a hypo-osmotic medium to a hyper-osmotic medium in elasmobranchs is initiated by an extra-cellular dehydration. Interestingly the response

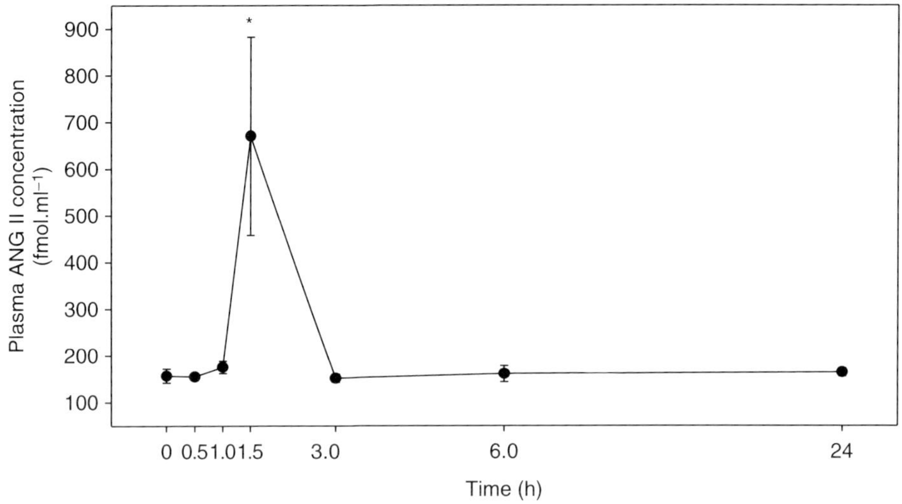

Figure 3. *Plasma levels of ANG II (fmol ml^{-1}) in* S. canicula *following rapid transfer from 80 to 100% SW* (n = 7). *Statistical comparison was made with values taken immediately after transfer. ANG II was significantly increased within 90 minutes of transfer. Values are expressed as a mean ± s.e.m.* (*p < 0.05) *(Students* t-*test).*

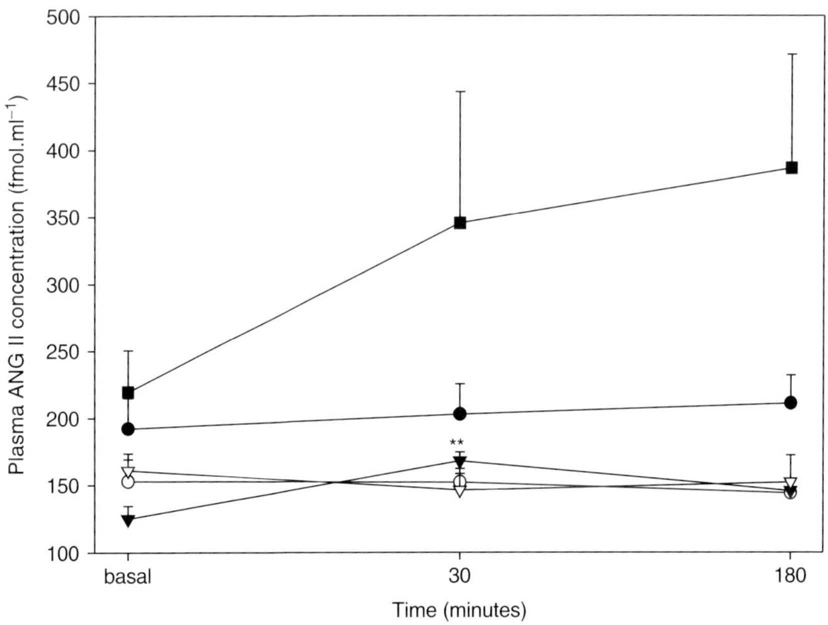

Figure 4. *Plasma levels of ANG II (fmol ml^{-1}) in* S. canicula *following: administration of 20% NaCl (●)* (n = 7); *2 M mannitol (○)* (n = 6); *75% sucrose (▼)* (n = 6); *vehicle (▽)* (n = 4); *and controlled haemorrhage (■)* (n = 9). *Statistical comparison was made with values taken immediately prior to the administration of test substances or controlled haemorrhage. Plasma ANG II was significantly greater 30 minutes post sucrose injection. Results are expressed as a mean ± s.e.m.* (**p < 0.01). *(Students* t-*test).*

following controlled haemorrhage was almost immediate in comparison to the delayed response following environmental transfer (Anderson *et al.*, 2002). It may be that marine elasmobranchs are required to reach a threshold of fluid loss prior to the increase in ANG II (*Figure 4*) and subsequent initiation of drinking.

5.4 *Cellular dehydration in elasmobranchs*

Cellular dehydration in 100% SW-acclimated *S. canicula* was induced by the intravascular administration of mannitol (2 M), 75% sucrose and 20% NaCl at a dose of 1 ml kg^{-1} via a coeliac arterial cannula, producing a theoretical cellular dehydration of 0.36, 0.33 and 0.47%, respectively (Anderson *et al.*, 2002). Vehicle (elasmobranch Ringer) was also administered at a similar dose. Following administration of 20% NaCl there was a significant decrease in drinking rate within 1 hour of administration.

The anti-dipsogenic effect of cellular dehydration induced after administration of hypertonic NaCl solutions was similar to that reported for FW- and SW-acclimated *A. japonica* (Takei *et al.*, 1988b), however, unlike the eel there was no observed increase in plasma ANG II. The release of renin is the primary step in the RAS cascade and is essential for an increase in plasma ANG II (Keeton and Campbell, 1981). In mammals and birds administration of hypertonic NaCl is thought to induce an increase in distal tubular sodium concentration in the nephron of the kidney which leads to the macula densa inhibiting renin release from the JG apparatus (Keeton and Campbell, 1981; Nishimura and Bailey, 1982; Takei *et al.*, 1988a). However, increased plasma osmolality induced by dextran or human serum albumin, but not NaCl, caused an increase in renin release in the unanaesthetized dog (Hall and Guyton, 1976). In the eel, administration of hypertonic NaCl and sucrose induced an increase in circulating levels of ANG II and plasma sodium concentration. The increase in plasma osmolality as a result of the administration of hypertonic solutions was thought to have induced an increase in renin release and a subsequent increase in plasma ANG II (Takei *et al.*, 1988b). This argument was supported by the evidence for the lack of a macula densa in eels (Sokabe and Ogawa, 1974). In elasmobranchs, however, histological evidence has demonstrated the presence of a macula densa (Lacy and Reale, 1989). Therefore, administration of hypertonic NaCl may serve to increase distal tubular sodium concentration and inhibit the release of renin, leading to the observed lack of increase in plasma ANG II levels (*Figure 4*).

In FW and SW eels, plasma sodium levels were significantly greater for up to 4 hours following injection of hypertonic saline (Takei *et al.*, 1988b). However, Anderson *et al.* (2002) showed no change in plasma osmolality and sodium after 30 minutes administration of 20% NaCl. Elasmobranchs are equipped with the extra-renal salt secreting gland, the rectal gland, which can clear large amounts of sodium chloride relatively quickly following an acute salt load (Shuttleworth, 1988). Hypervolaemia is known to induce an increase in rectal gland activity (Solomon *et al.*, 1985) and injection of 20% NaCl would induce an extra-cellular hyper-volaemia associated with cellular dehydration. In addition, a major stimulus for secretion in the rectal gland is C-type natriuretic peptide (CNP) (Anderson *et al.*, 1995; Solomon *et al.*, 1992) which is believed to be released in response to volume load (Loretz and Pollina, 2000). Interestingly, administration of hypertonic NaCl in eels induced a transient increase in plasma atrial natriuretic peptide (ANP), the major circulatory natriuretic peptide in eels (Kaiya and Takei, 1996). ANP is known to potently inhibit drinking in eels (Tsuchida and Takei, 1998). Perhaps significantly, in *S. canicula* CNP was shown to be 50 times more potent in its inhibition of drinking than the stimulatory effect of ANG II (Anderson *et al.*, 2001b). The release of CNP following cellular dehydration through administration of hypertonic NaCl would explain the inhibition of drinking in *S. canicula* and also the reported inhibition of drinking in eels despite the observed rise in circulating ANG II (Kaiya and Takei, 1996; Takei *et al.*, 1988b).

A similar result was observed in *S. canicula* where administration of hypertonic sucrose did not significantly increase drinking rate (Anderson *et al.*, 2002), despite the observed increase in circulating levels of ANG II within 30 minutes of injection of sucrose (*Figure 4*). Elasmobranchs would not be able to remove excess sucrose as rapidly as NaCl. Indeed it is unlikely that these fish are able to remove sucrose via the rectal gland so the kidney would be the sole route of sucrose excretion. Hence the increase in plasma levels of ANG II could be explained by an increase in distal tubular osmolality, from filtered sucrose, acting on the JG apparatus to increase renin secretion, as postulated for mammals and teleosts (Hall and Guyton, 1976; Takei *et al.*, 1988b). The increase in extra-cellular volume associated with an injection of hypertonic sucrose could also lead to an increase in circulating levels of CNP, stimulation of rectal gland secretion and inhibition of drinking. An increase in rectal gland secretion would lead to a decrease in plasma sodium concentration which was indeed observed after 30 minutes of administration of hypertonic sucrose (Anderson *et al.*, 2002). Furthermore, the observed increase in plasma ANG II following hypertonic sucrose injection was not mirrored by an increase in drinking which could be explained by an increase in circulating CNP levels in response to volume load.

Clearly there is a need to further investigate both renal and extra-renal function and the interaction of the peptides controlling these systems following administration of hypertonic solutions in order to further our understanding of the control of body fluid osmolality in elasmobranchs during cellular dehydration, or perhaps more importantly, extra-cellular hyper-volaemia.

6. Summary

In summary, it is evident that teleost and elasmobranch fish respond to extra-cellular dehydration by increasing drinking rate mediated by an increase in circulating levels of ANG II. However, although the primary stimulus for drinking may be the same, clearly the mechanisms involved in regulating ion and water balance are entirely different. In order to maintain ion and water balance in the face of cellular and extra-cellular dehydration, the integration and hormonal control of renal and extra-renal function in elasmobranchs has developed in a very different manner to that described for teleost fish.

Acknowledgements

The studies on *S. canicula* were carried out by W.G.A. with support from NERC grant GR3/11420 awarded to N.H. The studies on *T. scyllia* were carried out by N.H. as a visiting research Professor with support from the Ocean Research Institute, University of Tokyo, Japan. The authors would like to thank the Misaki Marine Biological Station, University of Tokyo, for the collection and maintenance of *T. scyllia*.

References

Anderson, W.G., Tierney, M.L., Takei, Y. and Hazon, N. (1995) Natriuretic hormones in elasmobranch fish; possible interactions with other endocrine systems. *Physiol. Zool.* **68:** 182.

Anderson, W.G., Cerra, M.C., Wells, A., Tierney, M.L., Tota, B., Takei, Y. and Hazon, N. (2001a) Angiotensin and angiotensin receptors in cartilagenous fishes. *Comp. Biochem. Physiol.* **128A:** 31–40.

Anderson, W.G., Takei, Y. and Hazon, N. (2001b) Possible interaction between the renin angiotensin system and natriuretic peptides on drinking in elasmobranch fish. In: *Perspectives in Comparative Endocrinology* (eds H.J.T. Goos, R.K. Rastogi, H. Vaudry and R. Perantoni). Monduzzi Editore, Bologna, Italy, pp. 753–758.

Anderson, W.G., Takei, Y. and Hazon, N. (2001c) The dipsogenic effect of the renin angiotensin system in elasmobranch fish. *Gen. Comp. Endocrinol.* **124:** 300–307.

Anderson, W.G., Takei, Y. and Hazon, N. (2002) Osmotic and volaemic effects on drinking rate in elasmobranch fish. *J. Exp. Biol.* **205:** 1115–1122.

Bean, J.W. (1942) Specificity in the renin–hypertensinogen reaction. *Am. J. Physiol.* **136:** 731–742.

Bernier, N.J., Gilmour, K.M., Takei, Y. and Perry, S.F. (1999) Cardiovascular control via angiotensin II and circulating catecholamines in the spiny dogfish, *Squalus acanthias. J. Comp. Physiol.* **169B:** 237–248.

Boylan, J.W. (1967) Gill permeability in elasmobranchs. In: *Sharks, Skates and Rays* (eds P.W. Gilbert, R. Mathewson and D.P. Rall). Johns Hopkins University Press, Baltimore, pp. 197–206.

Boylan, J.W. (1972) A model of passive reabsorption in the elasmobranch kidney. *Comp. Biochem. Physiol.* **42A:** 27–30.

Burger, J.W. (1962) Further studies on the function of the rectal gland in the spiny dogfish. *Physiol. Zool.* **38:** 191–196.

Burger, J.W. and Hess, W.N. (1960) Function of the rectal gland in the spiny dogfish. *Science.* **131:** 670–671.

Conlon, J.M. (1999) Bradykinin and its receptors in non-mammalian vertebrates. *Regul. Pept.* **79:** 1–81.

Doyle, W.L. and Gorecki, D. (1961) The so-called chloride cells of the fish gill. *Physiol. Zool.* **34:** 81–85.

Fitzsimons, J.T. (1979) *The Physiology of Thirst and Sodium Appetite* (ed. J.T. Fitzsimons). Cambridge University Press, Cambridge, pp. 1–121.

Gardiner, S.M., Kemp, P.A. and Bennett, T. (1993) Differential effects of captopril regional haemodynamic responses to angiotensin I and bradykinin in conscious rats. *Br. J. Pharmacol.* **108B:** 169–175.

Hall, J.E. and Guyton, A.C. (1976) Changes in renal hemodynamics and renin release caused by increased plasma oncotic pressure. *Am. J. Physiol.* **231:** 1550–1556.

Hazon, N., Balment, R.J., Perrott, M. and O'Toole, L.B. (1989) The renin–angiotensin system and vascular and dipsogenic regulation in elasmobranchs. *Gen. Comp. Endocrinol.* **74:** 230–236.

Hazon, N., Tierney, M.L., Anderson, W.G., Mackenzie, S., Cutler, C. and Cramb, G. (1997) Ion and water balance in elasmobranch fish. In: *Ionic Regulation in Animals* (eds N. Hazon, F.B. Eddy and G. Flik). Springer, Heidleberg, pp. 70–86.

Hazon, N., Tierney, M.L. and Takei, Y. (1999) Renin–angiotensin system in elasmobranch fish: a review. *J. Exp. Zool.* **284:** 526–534.

Hentschel, H. (1987) Renal architecture of the dogfish *Scyliorhinus caniculus* (Chondrichthyes, Elasmobranchii). *Zoomorph.* **107:** 115–125.

Hirano, T. (1974) Some factors regulating water intake in the Japanese eel, *Anguilla japonica. J. Exp. Biol.* **61:** 737–747.

Jampol, L.M. and Epstein, F.M. (1970) Sodium-potassium-activated adenosine triphosphatase and osmotic regulation by fishes. *Am. J. Physiol.* **218:** 607–611.

Joss, J.M.P., Itahara, Y., Watanabe, T.X., Nakajima, K. and Takei, Y. (1999) Teleost-type angiotensin is present in the Australian lungfish. *Gen. Comp. Endocrinol.* **114:** 206–212.

Kaiya, H. and Takei, Y. (1996) Osmotic and volemic regulation of atrial and ventricular natriuretic peptide secretion in conscious eels. *J. Endocrinol.* **149:** 441–447.

Keeton, T.K. and Campbell, W.B. (1981) The pharmacologic alteration of renin release. *Pharmacol. Rev.* **32:** 81–227.

Kent, B. and Olson, K.R. (1982) Blood flow in the rectal gland of *Squalus acanthias. Am. J. Physiol.* **243:** 296–303.

Kobayashi, H. and Takei, Y. (1996) *The renin angiotensin system, comparative aspects* (eds H. Kobayashi and Y. Takei). Zoophysiology, Vol. 35. Springer, Berlin, Heidleberg.

Kobayashi, H., Uemura, H., Takei, Y., Itatsu, N., Ozawa, M. and Ichinohe, K. (1983) Drinking induced by angiotensin II in fishes. *Gen. Comp. Endocrinol.* **49:** 295–306.

Lacy, E.R. and Reale, E. (1989) Granulated peripolar epithelial cells in the renal corpuscle of marine elasmobranch fish. *Cell Tiss. Res.* **257:** 61–67.

Lacy, E.R. and Reale, E. (1990) The presence of a juxtaglomerular apparatus in elasmobranch fish. *Anat. Embryol.* **182:** 249–262.

Lacy, E.R. and Reale, E. (1995) Functional morphology of the elasmobranch nephron and retention of urea. In: *Cellular and molecular approaches to fish ionic regulation* (eds C.M. Wood and T.J. Shuttleworth). Academic Press, New York, pp. 107–146.

Lacy, E.R., Reale, E., Schlusselburg, D.S., Smith, W.K. and Woodward, D.J. (1985) A renal countercurrent system in marine elasmobranch fish: a computer assisted reconstruction. *Science.* **227:** 1351–1354.

Laurent, P. and Dunel, S. (1980) Morphology of gill epithelia in fish. *Am. J. Physiol.* **238:** R147–R159.

Loretz, C.A. and Pollina, C. (2000) Natriuretic peptides in fish physiology. *Comp. Biochem. Physiol.* **125B:** 169–187.

Nishimura, H. (1985) Evolution of the renin angiotensin system and its role in control of cardiovascular function in fishes. In: *Evolutionary biology of primitive fishes* (eds R.E. Foreman, A. Gorbman, J.M. Dodd and R. Olsson). Plenum Press, New York, pp. 275–293.

Nishimura, H. and Bailey, J.R. (1982) Intrarenal renin angiotensin system in primitive vertebrates. *Kidney Int.* **22:** S185–S192.

Nishimura, H., Oguri, M., Ogawa, M., Sokabe, H. and Imai, H. (1970) Absence of renin in the kidney of elasmobranchs and cyclostomes. *Am. J. Physiol.* **218:** 911–915.

Nishimura, H., Lunde, L.G. and Zucker, A. (1979) Renin response to haemorrhage and hypotension in the aglomerular toadfish, *Opsanus tau. Am. J. Physiol.* **237:** H105–H111.

Okawara, Y., Karakida, T., Aihara, M., Yamaguchi, K. and Kobayashi, H. (1987) Involvement of angiotensin II in water intake in the Japanese eel, *Anguilla japonica. Zool. Sci.* **4:** 523–528.

Schmidt-Nielsen, K. (1997) Water and osmotic regulation. In: *Animal Physiology*, 5th edn (ed. K. Schmidt-Nielsen). Cambridge University Press, Cambridge, pp. 301–391.

Shuttleworth, T.J. (1988) Salt and water balance. In: *The Physiology of Elasmobranch Fishes* (ed. T.J. Shuttleworth). Springer, London, pp. 171–199.

Sokabe, H. and Ogawa, M. (1974) Comparative aspects of the juxtaglomerular apparatus. *Int. Rev. Cytol.* **37:** 271–327.

Smith, C.P. and Wright, P.A. (1999) Molecular characterization of an elasmobranch urea transporter. *Am. J. Physiol.* **276:** R622–R626.

Smith, H.W. (1936) The retention and physiological role of urea in the Elasmobranchii. *Biol. Rev.* **11:** 49–82.

Solomon, R., Taylor, M., Sheth, S., Silva, P. and Epstein, F.H. (1985) Primary role of volume expansion in stimulation of rectal gland function. *Am. J. Physiol.* **248:** R638–R640.

Solomon, R., Protter, A., McEnroe, G., Porter, J.G. and Silva, P. (1992) C-type natriuretic peptides stimulate chloride secretion in the rectal gland of *Squalus acanthias*. *Am. J. Physiol.* **262:** R707–R711.

Takei, Y. (2000) Comparative physiology of body fluid regulation in vertebrates with special reference to thirst regulation. *Jap. J. Physiol.* **50:** 171–186.

Takei, Y., Okawara, Y. and Kobayashi, H. (1988a) Control of drinking in birds. In: *Progress in Avian Osmoregulation* (eds M.R. Hughes and A.C. Chadwick). Leeds Philosophical and Literary Society Ltd, Leeds.

Takei, Y., Okubo, J. and Yamaguchi, K. (1988b) Effects of cellular dehydration on drinking and plasma angiotensin II level in the eel, *Anguilla japonica*. *Zool. Sci.* **5:** 43–51.

Takei, Y., Hasegawa, Y., Watanabe, T.X., Nakajima, K. and Hazon, N. (1993) A novel angiotensin I isolated from an elasmobranch fish. *J. Endocrinol.* **139:** 281–285.

Takei, Y., Itahara, Y., Butler, D.G., Watanabe, T.X. and Oudit, G.Y. (1998a) Tetrapod-type [Asp[1]] angiotensin is present in a holostean fish, *Amia calva*. *Gen. Comp. Endocrinol.* **110:** 140–146.

Takei, Y., Tsuchida, T. and Tanakadate, A. (1998b) Evaluation of water intake in seawater adaptation in eel using synchronised drop counter and pulse injector system. *Zool. Sci.* **15:** 677–683.

Takei, Y., Tsuchida, T., Li, Z. and Conlon, J.M. (2001) Antidipsogenic effects of eel bradykinins in the eel, *Anguilla japonica*. *Am. J. Physiol.* **281:** R1090–R1096.

Tierney, M.L., Luke, G., Cramb, G. and Hazon, N. (1995) The role of the renin–angiotensin system in the control of blood pressure and drinking in the European eel, *Anguilla anguilla*. *Gen. Comp. Endocrinol.* **100:** 39–48.

Tierney, M.L., Takei, Y. and Hazon, N. (1998) A radioimmunoassay for the determination of angiotensin II in elasmobranch fish. *Gen. Comp. Endocrinol.* **111:** 299–305.

Tsuchida, T. and Takei, Y. (1998) Effects of homologous atrial natriuretic peptide on drinking and plasma ANG II levels in eels. *Am. J. Physiol.* **275:** R1605–R1610.

Tsuchida, T. and Takei, Y. (1999) A potent dipsogenic action of homologous angiotensin II infused at physiological doses in eels. *Zool. Sci.* **16:** 479–483.

Yamaguchi, K. (1981) Effects of water deprivation on immunoreactive angiotensin II levels in plasma, cerebroventricular perfusate and hypothalamus of the rat. *Acta Endocrinol.* **97:** 137–144.

Yancey, P.H. and Somero, G.N. (1980) Methylamine osmoregulatory solutes of elasmobranch fishes counteract urea inhibition of enzymes. *J. Exp. Zool.* **212:** 205–213.

3

Drinking in (very young) fish

G. Flik, S. Varsamos, P.M.G. Guerreiro, X. Fuentes, M.O. Huising and J.C. Fenwick

1. Introduction

At the start of this rather topical review on drinking in (very young) fish, we want to give credit to an excellent, comprehensive and recent review on drinking in fish, *viz.* that by Fuentes and Eddy (1997); the latter paper serves as the ideal first reading for those not yet familiar with or starting the study of the physiology of drinking and its consequences in fish.

Fish live in almost every aquatic (and beyond) niche on earth, ranging from waters that are extremely soft and mineral-poor to waters that contain ion concentrations far exceeding those found in their body fluids. In brackish waters conditions for hydromineral homeostasis are favourable in the sense that water and ion movements between the fish and the ambient medium are minimal as the mineral content of the blood plasma compartment of the fish resembles the ambient medium to a larger degree and osmosis- and diffusion-related water and ion movements are thus limited (Vonck *et al.*, 1998). However, fish show hydromineral homeostasis essentially independently of the environment as other vertebrates do and thus in fresh water (FW) fish must actively remove water from their body and take up ions, in sea-water (SW) they must gain water and remove ions. Fish in a SW environment drink and eliminate excess ions via branchial, renal and intestinal routes. In FW drinking should be avoided or limited, for instance just to moisten the food taken in (Ruohonen *et al.*, 1997; fish have no salivary glands). However, neural mechanisms may be activated and hormones that initiate drinking may be released under stress conditions and invoke compulsory drinking in a FW environment. We have measured drinking rates in very young fish using inert volume tracer to assess the onset of drinking and its consequence for the hydromineral balance of the fish and we evaluated the role of the mineral content of the ambient medium in drinking. Hormones that influence calcium metabolism of the fish (stanniocalcin and parathyroid hormone related protein, PTHrP) influence drinking, and may do so indirectly through vasodilatory or vasoconstrictive actions and effects on the renin angiotensin system(s) (RAS) of the fish. Drinking induced by mild stress in FW fish could have important applications in disease treatment (vaccination) as the intestine

Osmoregulation and Drinking in Vertebrates, edited by N. Hazon & G. Flik.
© 2002 BIOS Scientific Publishers Ltd, Oxford

represents a more leaky (also for antigens) epithelium (e.g. Ellis, 1998) compared to the tight epithelia of skin and gills that are in direct contact with the water (Santos *et al.*, 2001). Along the same line of thought, environmentally- or stress-induced drinking will have consequences for gastrointestinal actions of waterborne toxic substances. Understanding drinking in very young fish may thus shed light on their vulnerability to disturbances and will have important implications for aquaculture.

Fish—just as all other vertebrates do—have realized a sophisticated degree of homeostasis of the body fluids, i.e. they keep their body fluids more or less constant in ionic and water contents and the values found for minerals in blood plasma of fish are never drastically out of line with the values seen in a terrestrial vertebrate. SW fish drink to compensate for osmotic water loss; to counteract hypertonicity of the body fluids as a result of inward diffusion of ions directly from the water over the (non-keratinizing) body surfaces and the salt water drunk, they actively secrete ions mainly via their gills that are equipped with specialized ion transporting cells. Thus, the SW drunk (SW contains for instance around 500 mM NaCl, 10 mM Ca, 60 mM $MgSO_4$) is taken in for the water, not for the ions, and to this end the fluid ingested is desalinated in the oesophagus to allow for water absorption in more distal parts of the intestinal tract via $Na^+/K^+/Cl^-$-cotransporter-driven pathways (Musch *et al.*, 1982). Importantly, drinking in SW may thus, indirectly, depend on integumental or branchial chloride cells and their regulatory machinery needed to remove from the body the excess ions taken 'accidentally' in for water needs. If we assume that the first fish originated in a SW/hyper-ionic environment, the development or possession of drinking mechanisms as we now know must have been of ultimate importance to internal homeostasis and survival. Consequently, the mechanisms for drinking and their regulation must be very old and an already perfected invention of the first fish emerging 500 to 450 million years ago in the seas on this planet. At the same time—and one cannot imagine that there have ever been life stages in which the interdependent physiological key mechanisms as we study them now were less-developed—the early fish had to be equipped with mechanisms for secretion and/or excretion of mono- and divalent ions. Indeed an overload of ions is taken up (or intrudes the body), is used for manipulation of water flow and the realization of the ionic disequilibria characteristic of a living cell, organ and organism, but any surplus needs to be removed. Chloride cells play a crucial role in this process (Zadunaiski, 1997).

The question arises at what developmental stage of the fish drinking first occurs. If one assumes that early developmental stages of fish harbour phylogenetically old traits, maybe even traits inherited from the immediate predecessors of the vertebrates, one could study in fish larvae the early onset of drinking in vertebrates. Halibut larvae start drinking before the anal opening becomes functional, in herring larvae before functional gills are present (Tytler and Blaxter, 1988a, b) and drinking seems to be an absolute requirement as soon as the fish hatch, i.e. when they are no longer protected by the egg membrane that is notoriously impermeable to water and ions (Potts and Eddy, 1973).

2. The onset of drinking

Proceeding from the reasoning above that drinking is a requirement for fish in SW, one should study a SW species. There are two particular reasons then to consider the sea bass (*Dicentrarchus labrax*) as an intriguing model to study. First it is a SW species with an exceptional broad range in salt tolerance. The euryhalinity of the fish offers the advantage that the fish can be studied also in hypo-ionic medium, where one may predict that drinking is no longer a requirement. Moreover, this species goes through a more or less stepwise metamorphosis in early development where the osmoregulatory capacity develops and becomes defined (Barnabé *et al.*, 1976).

Consistent drinking (35 nl h^{-1}) was observed in larvae of around 5 mm (0.55 mg), i.e. when the sea bass first opens its mouth and starts exogenous feeding (earlier stages were as yet not included in our studies). Longitudinal sections of paraffin-embedded specimens of these stages show that the intestinal tract also shows an anal opening around this time (S. Varsamos, unpublished) suggesting a tract fully prepared for feeding and drinking. As mentioned above, halibut larvae start drinking before anal opening and thus water processing for osmoregulatory purposes may not depend on an open intestinal tract; indeed consensus exists that most of the SW drunk by SW fish is processed and absorbed in early segments of the intestinal tract. Transfer of these larvae to a hypo-ionic medium (5% SW) slightly decreased drinking rate (32 nl h^{-1}); transfer to hypersaline water resulted in significant mortality and, therefore, we have excluded this one group from further considerations. In larvae of 15 mm (around 100 mg), when the fish are at the end of larval development, drinking has increased to 1.7 μl $hour^{-1}$ and transfer to lower salinity decreases drinking immediately; transfer to higher salinity immediately enhances drinking. This trend becomes very much more pronounced in the juvenile stages (*Figure 1*). One could argue that the drinking rate estimated for the youngest stages is merely due to adsorption of tracer to the body wall (we measured the whole larva, not the isolated gut) and thus the rate an experimental artefact. However, similar rates were consistently reported, for instance for very young halibut, herring, turbot and sea bream larvae (Tytler and Blaxter, 1988a, b; Tytler and Ireland, 1994; Reitan *et al.*, 1998; Guerreiro *et al.*, 2001) and thus may be considered realistic. Moreover, transfer of the sea bass larvae to 5% SW slightly but consistently decreased drinking (and in these experiments the radioactive concentration was kept constant in all conditions), indicative of a predictable regulatory response, as the larvae would need to decrease or even stop drinking in such medium. However, in all developmental stages of the sea bass studied significant drinking was observed in 5% SW and this may indicate that handling of fish (which was the same for all fish) as occurred in these experiments may have induced significant drinking. Turbot larvae adapted to 50% SW instead of full-strength SW also drank more than anticipated (Brown and Tytler, 1993); maybe mechanisms that down-regulate drinking in adults are not yet fully developed in such young larvae? The drinking rate (DR, in nl h^{-1}) of fish in 25% SW increases linearly with weight (W, in mg) of the fish ($DR = 1.8\ W + 64.5$ nl h^{-1}; $r^2 = 0.9985$, $P < 0.001$). This simple relationship further corroborates the notion that our estimates of drinking in very young stages may be realistic. Moreover, very early larval stages of the sea bass (Varsamos *et al.*, 2002) as well as other fish such as tilapia (Heijden *et al.*, 1999a) are equipped with an abundance of functional chloride cells in the integumental epithelia of skin, yolk sac and developing gills and thus the ionoregulatory consequences of drinking SW, *i.a.* NaCl excretion by chloride cells, can be dealt with by the larva. Indeed, the total number of chloride cells in FW tilapia larvae, that do not drink or drink very little, is significantly lower than in SW larvae (Heijden *et al.*, 1999a) and this lower number may relate both to a generally

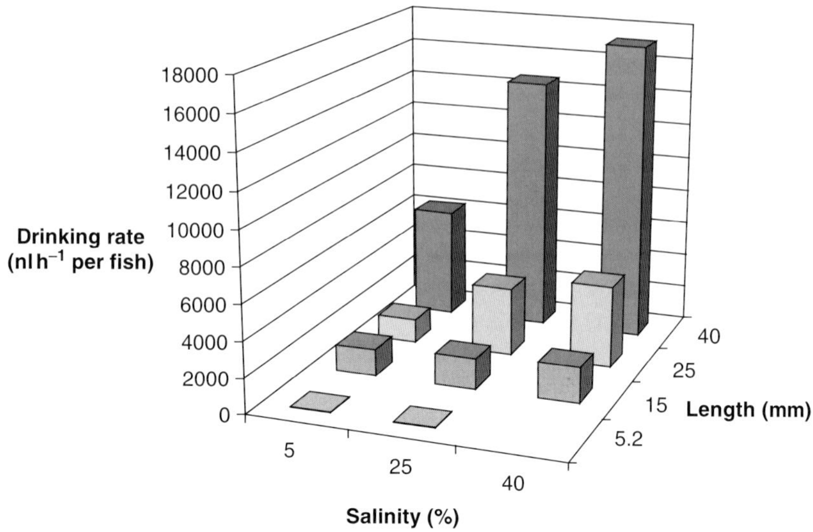

Figure 1. *Drinking in very young sea bass (*Dicentrarchus labrax*) raised in 25% SW. ^{51}Cr-EDTA (up to 370 kBq ml^{-1}) was added to the water (17°C) as inert marker of water volume. In adult fish it was confirmed that ^{51}Cr-EDTA remains confined to the gastrointestinal tract and does not appear in the plasma compartment upon exposure of the fish to waterborne ^{51}Cr-EDTA for 4 to 6 h. Larvae (40 per 100 ml) and juveniles (20 per 2 l) were exposed for up to 4 h to the tracer, then rinsed three times in tracer-free medium of the same ionic composition and subsequently euthanized in phenoxyethanol (0.5%) and digested in 0.25–1.0 ml H_2O_2 at 60°C for 4 h. In all instances filtered (0.2 µm) water was used. To the digests (and water samples) 3 ml HiSafe scintillation cocktail was added and the radioactivity was determined in a Wallac 1410™ Liquid Scintillation Analyser. The drinking rate was calculated in nl h^{-1} per fish on the basis of linear uptake of tracer. In a separate series of experiments it was assessed that the uptake of tracer was linear for up to 4 h in larvae of around 5 mm, up to 2 h in larvae of 15 mm and juveniles of 25 mm and for 1 h in juveniles of 45 mm. Fish were transferred by netting to the respective media (5%, 25% and 40% SW) at the very start of the experiment.*

lower requirement for ion transport in FW as well as to a reduced pressure on ion regulation in fish that drink little or none. We would conclude then that a sea bass starts drinking and can regulate this process in very early stages of development, a likely adaptation of the species to cope with salinity fluctuations in coastal waters where its early development normally takes place. This notion may further have important implications for aquaculture practices.

3. Calciotropic hormones and drinking in fish

In teleost fish we now know at least four protein hormones that exert calciotropic effects: prolactin, stanniocalcin, parathyroid hormone related protein (PTHrP) and calcitonin.

3.1 *Prolactin*

Prolactin is considered a hormone of prime importance for survival of fish in a FW enviroment (Bern, 1975). The hormone exerts a major control over integumental epithelia, making them more tight, i.e. impermeable to the movement of water and ions (including calcium ions), a requisite for survival of a vertebrate with a non-keratinizing integument (and depending on hydromineral homeostasis of the body fluids) in a hypo-osmotic and hypo-ionic environment. Being a typical pleiotropic hormone, one could list pages of actions and functions for prolactin, but a peculiar action to mention is its calciotropic action in FW fish: prolactin injected in FW tilapia induces hypercalcemia, and it enhances chloride cell activity and the calcium transporting enzymes in these cells, that warrant the uptake of calcium from the water (Flik *et al.*, 1984, 1986, 1989, 1994). Through these actions prolactin supports survival of fish in FW and more so when the water is poorer in calcium ions and the integument may become more leaky. Ambient calcium levels are inversely related to prolactin cell activity (Wendelaar Bonga *et al.*, 1985). Indeed, when euryhaline species such as tilapia (*Oreochromis mossambicus*) are transferred from FW to SW (containing roughly 10 mmol l^{-1} Ca) prolactin cells become almost silent and plasma prolactin levels almost undetectable. This permeability control, hypercalcemic actions and an inhibition of drinking would all fit in a FW adaptating role for prolactin and thus argue against a role for prolactin, under normal circumstances, in the direct regulation of drinking. Comparing the conditions of gills and gut in FW and SW conditions and the calcium transport in these organs (*Table 1*), again no direct link between prolactin cell activity and drinking is found. The increased leakiness of gills and gut in SW fish is in accordance with a lack of permeability control by prolactin; the increased leakiness of the intestinal epithelium in SW resulting in zero net flux of calcium (Schoenmakers *et al.*, 1993) is a perfect adaptation to the drinking of SW: tilapia takes up more calcium from the water via its gills than is required for growth-related accumulation of the element both in FW and SW (Vonck *et al.*, 1998), so why then take up even more via the gut if drinking of SW becomes compulsory for osmotic reasons?

Interestingly though, when blood is withdrawn from tilapia and a significant hypo-volaemia induced, the fish start drinking in FW (!) and plasma prolactin levels rise (see Chapter 7, Hirano *et al.*, 2002; this volume). It is at present assumed that hypo-volaemia is the strongest stimulus for drinking in fish (Takei, 2000) and induces an angiotensin-II surge that evokes both the drinking and the prolactin response. Whatever the principle at work under these conditions, the results warrant research into a so far unexpected role for prolactin in (solving the consequences of) drinking and the regulation of this behaviour. When fish hatch and lose the protective egg membrane, prolactin will be crucial in a FW species to control integumental

Table 1. *A comparison of freshwater (FW) and sea-water (SW) tilapia*

Water	DR	Prl cell	Gills	Ca^{2+} influx	Gut	$J_{ms}(Ca^{2+})$	$J_{sm}(Ca^{2+})$	$J_{net}(Ca^{2+})$
FW	0–10	Active	Tight	3.5 ± 0.6	Leaky	27.2 ± 8.5	19.6 ± 7.5	17.1 ± 8.9
SW	75	Inactive	Leaky	4.4 ± 0.4	More leaky	26.5 ± 6.7	28.6 ± 9.9	n.s.

Drinking rate (DR) is given in μl h^{-1} per 100 g fish; branchial Ca^{2+} influx in μmol h^{-1} per 100 g fish; Ca^{2+} fluxes across (stripped) intestinal epithelium are in nmol h^{-1} cm^{-2}. Prl cell: prolactin cell; J_{ms}: mucosa to serosa flux; J_{sm}: serosa to mucosa flux; $J_{net} = J_{ms} - J_{sm}$; n.s.: no significant net flux. Data on branchial calcium uptake and intestinal calcium transport taken from Flik *et al.*, 1986 and Schoenmakers *et al.*, 1993.

permeability to water and ions and maybe in control of drinking as well. Larval stages of fish deserve more attention in this respect in research on the regulation of drinking. Considering both external (e.g. ambient Ca^{2+} or osmolarity) and internal (various endocrines) stimuli and their potential interplay and consequences may open our understanding of the role of calciotropic hormones in drinking.

3.2 *Stanniocalcin*

Another example of a suggestive, yet unclear, relation between drinking and a calciotropic hormone comes from our studies on the actions of stanniocalcin in (adult) eels. In these fish it is rather feasible to remove the prime source of stanniocalcin: in the eel two corpuscles are always found in the posterior kidney, often located on the ventral surface and the glands are easily removed. Successful extirpation of the glands is followed by a phenomenal hypercalcemia (Hanssen *et al.*, 1989; *Table 2*). Both total and ionic calcium go up drastically, provided there is significant calcium in the water (Flik, 1990), resulting from an unleashed calcium influx via the gills (Verbost *et al.*, 1993; Heijden *et al.*, 1999b).

We were primarily interested in the whole body $^{45}Ca^{2+}$ fluxes in these fish; since we included SW eels in these experiments, we first determined the $^{45}Ca^{2+}$ ingested during the experiment on the basis of ^{45}Ca content of the intestinal tract so as to correct whole body uptake of $^{45}Ca^{2+}$ for intestinal uptake (Heijden *et al.*, 1999b). To our surprise, when these data are transformed to drinking rate (assuming ^{45}Ca entered the gut only via drinking, and did not leave the intestine to a significant degree; Pang *et al.*, 1980), drinking was very significantly enhanced both in FW and SW eels once the stanniocalcin source was removed. These results could be reproduced closely using ^{51}Cr-EDTA as tracer for drinking a similar set-up (*Figure 2*).

The question arises whether the removal of stanniocalcin forms a stimulus for drinking and thus whether stanniocalcin normally inhibits drinking. For long a debate has been going on concerning the bioactive components of the corpuscles of Stannius. The corpuscles may produce bioactive principles other than stanniocalcin, *viz.* teleocalcin (Copp *et al.*, 1985) and even as yet unidentified vasoactive factors (Chester Jones *et al.*, 1966; Butler and Oudit, 1995). As eels become somewhat

Table 2. *Effects of sham operation (Sham) or stanniectomy (STX) on plasma sodium, total and ionic calcium and magnesium and urinary calcium concentrations of eels adapted to FW or SW*

Treatment		Plasma Na	Plasma Ca	Plasma Ca^{2+}	Plasma Mg	Urine Ca
FW	Sham	150 ± 10	2.7 ± 0.3	1.25 ± 0.15	1.4 ± 0.1	0.6 ± 0.4
	STX	121 ± 12*	6.5 ± 1.8*	3.15 ± 0.35*	1.4 ± 0.1	4.2 ± 1.2*
SW	Sham	153 ± 9	3.2 ± 0.4	1.28 ± 0.24	1.5 ± 0.2	6.9 ± 1.5
	STX	147 ± 6	7.7 ± 1.2*	3.55 ± 0.65*	1.7 ± 0.4	7.1 ± 1.5

All values are in mmol l^{-1} and based on at least six fish each. Removal of the source of stanniocalcin has hyponatremic effects in FW eel but not in SW eel, induces a strong and specific hypercalcemia; renal pathways seem to be involved in excretion of part of the surplus of calcium in FW eels. *$P < 0.01$, compared to sham controls. Part of the data were taken from Heijden *et al.*, 1999b and Hanssen *et al.*, 1989.

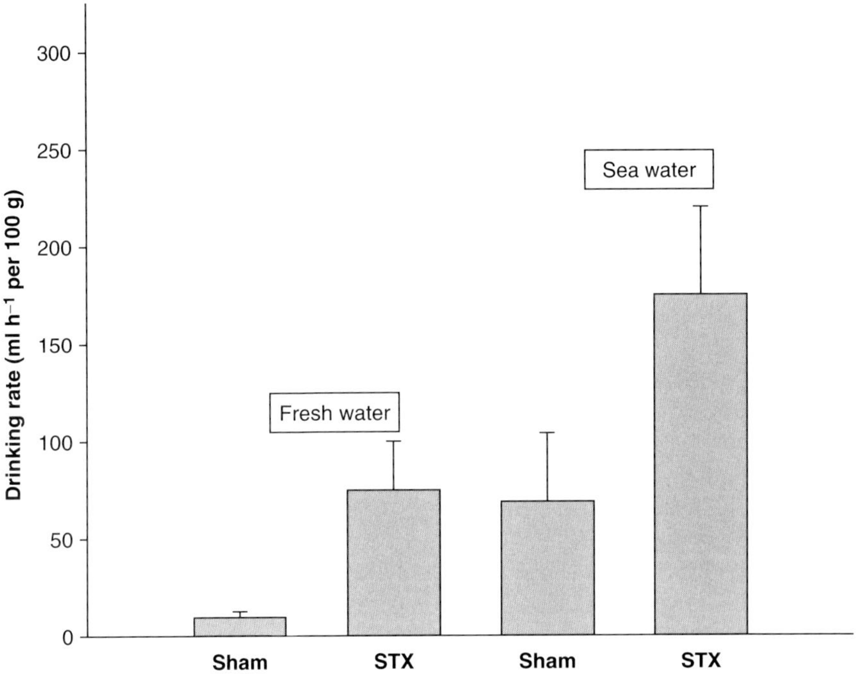

Figure 2. *Drinking in sham-operated and stanniectomized European eels (*Anguilla anguilla*) acclimatized for 4 weeks to FW or artificial SW (32%). Animals were allowed to recover from surgery for two weeks, before the start of the experiments. Drinking rate was estimated on the basis of gut ^{51}Cr-EDTA content after 1 h exposure in the water; upon completion of the experiment the fish were euthanized in phenoxyethanol (1%) and, belly down, snap-frozen on solid CO_2. Once frozen the complete frozen intestinal tract was removed and digested in H_2O_2 and aliquots were analysed for tracer content. Drinking rate was calculated on the basis of calculated gut total counts and water radioactive concentration for ^{51}Cr. Sham: sham-surgery; STX: fish with corpuscles of Stannius removed for at least 2 weeks.*

lethargic and hypotensive after stanniectomy (Butler and Oudit, 1995), the most likely explanation for the drinking seen after stanniectomy is an activation of the RAS in response to the hypotension, followed by drinking (Tierney *et al.*, 1995; Tsuchida and Takei, 1999). In line with this conclusion is our observation (P. Guerreiro and G. Flik, unpublished) that drinking in the SW-adapted sea bream is not influenced by injecting the fish with a stanniocalcin antiserum, a treatment that induces mild hypercalcemia and thus likely and specifically eliminated (immunoneutralized) at least a significant part of the circulating stanniocalcin (Fenwick *et al.*, 1995). Unfortunately, data on vasoactive effects of this treatment are as yet lacking. The search for the actual vasoactive principle of the corpuscles seems indicated. Strangely enough, to the best of our knowledge the obvious experiment, i.e. an analysis of the effects of stanniocalcin treatment of drinking SW eels, has not been performed so far, but it may shed definitive light on this intriguing topic. As stanniectomy also induces enhanced tubular reabsorption of water in eel kidney (Butler and Alia Cadinouche, 1995), aquaporins as mediators of such water movement must be considered a potential target of stanniocalcin.

3.3 *Parathyroid hormone related protein*

When sea bream (*Sparus aurata*) larvae, kept in full-strength SW (36%), are exposed to PTHrP ($PTHrP_{1-38}$ from *Fugu rubripes*) in the water, they enhance their uptake of calcium from the water, reduce the loss of calcium to the water and diminish drinking in a PTHrP concentration-dependent manner; in dilute SW (12%) similar effects are observed on calcium uptake, but drinking is not influenced (Guerreiro *et al.*, 2001; *Figure 3*). The reduction of drinking in SW was interpreted as a measure to avoid an excessive salt and calcium load, particularly a problem in a SW environment.

In dilute SW (larvae were reared in the media mentioned, no rapid transfers as described above under **2** for sea bass larvae), the sea bream larvae (continue to) drink, albeit at about 50% of the rate seen in SW, even though the medium (12%, around 380 mOsmol kg H_2O^{-1}) has an osmolality presumably close to that of the body fluids. The larvae in this experiment were 50 days post hatching, so inhibition of drinking in response to a diluted ambient medium appears (still) absent at this stage in sea bream. The lower drinking rate may have caused a smaller uptake of PTHrP from the water and consequently lower plasma levels of PTHrP (estimated to 12–200 pM in larvae exposed to 0.1–10 nM PTHrP in SW); but if they drink at half the rate in dilute SW and if we may extrapolate the data on the estimates of plasma PTHrP in SW larvae to larvae in dilute SW, one would still predict physiologically relevant plasma values of 15–100 pM in the dilute SW larvae. Then at least at the highest concentration of PTHrP in dilute SW larvae may be predicted to affect drinking but this was not seen. Maybe the calcemic status differed among the groups, or was

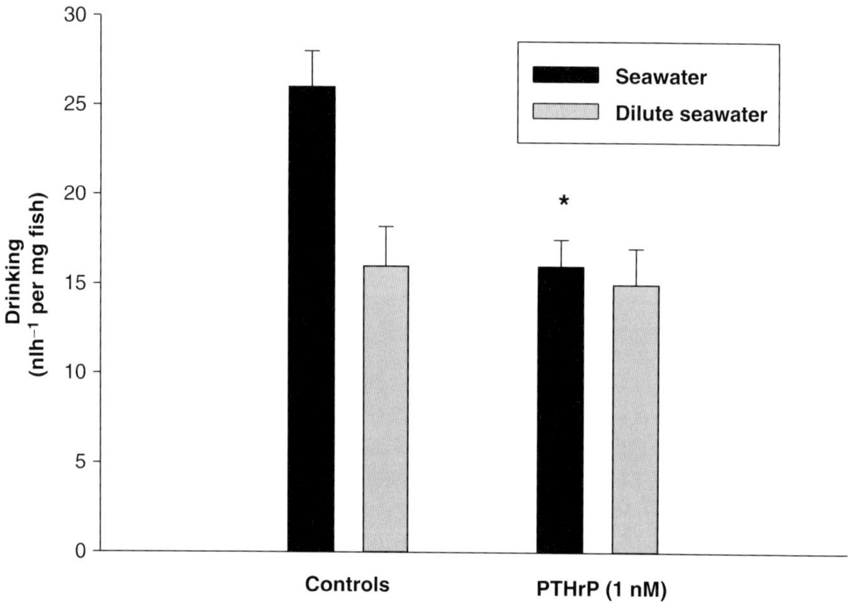

Figure 3. *Drinking in sea bream (*Sparus aurata *L.) larvae (30–60 mg) exposed during a 4-h period to 1 nM PTHrP1-38 in the water. Larvae were adapted to either full-strength SW (36%, black bars) or dilute SW (12%, grey bars). * Significantly different from controls (P<0.01; mean values; error bars: SE;* n *= 20–25). Drinking is significantly inhibited in SW but not in dilute SW. Data taken in a modified format from Guerreiro* et al.*, 2001.*

the interaction with, or tonus of any of the other endocrines involved in the adaptation to these vastly different media the cause of this difference? PTHrP is known to stimulate centrally the release of arginine vasopressin in the rat, resulting in anti-diuresis and pressor effects (Yamamoto *et al.*, 1998) and these effects could of course influence the regulation of drinking. In euryhaline species the pituitary content of arginine vasotocin (the fish equivalent of AVP) responds to the osmotic challenges associated with transfer of the fish between FW and SW (Balment *et al.*, 1993); in trout vasotocin levels are lowest in dilute SW (1/3 SW) and drastically higher in SW fish (Pierson *et al.*, 1995), and thus differences in PTHrP targets (if this holds true for fish as for rats) may exist. Interestingly, in eels transitory rises in AVT are accompanied by transitory rises in angiotensin II (ANG II) (Balment *et al.*, 1993) and thus more links with endocrine interactions in regulation of drinking are all too obvious. The presence of PTHrP in brain and pituitary tissues of fish (Ingleton and Danks, 1996) makes it highly likely that this hormone can influence directly and indirectly the regulatory systems that control drinking, a vast and unexplored area of research.

Another intriguing observation in these experiments is of course that antigen is taken up by the larvae, likely by drinking it, without apparent loss of bioactivity even when the antigen is a protein. This notion has important consequences for aquaculture practices. Other examples of antigen/protein uptake will be given in section 4 of this paper.

3.4 *Calcitonin*

Calcitonin is generally accepted as a hypocalcemic hormone, and indeed the first sources of calcitonin for clinical use were of fish origin (salmon). The long half-life in humans of the heterologous preparation and low antigenicity mean that, even now with bioengineered homologous preparations available, salmon calcitonin is still used in the clinic (Azria *et al.*, 1995). Hypocalcemic actions of calcitonin in fish are very mild (Wendelaar Bonga, 1980), or often not evident (Wendelaar Bonga and Lammers, 1982) and do not compare to the drastic hypocalcemic effects seen in fish by administration of stanniocalcin (Hanssen *et al.*, 1989). This situation has cast doubts over the role of calcitonin in calcium physiology in fish. Very little is known too about calcitonin, or its gene-related peptide, in relation to drinking in fish. Also for mammals literature on this topic is scant. However, from studies on rats (Schmid *et al.*, 1998) it is known that calcitonin injected subcutaneously may stimulate drinking (in a way ANG II does) via an action on subfornical neurons. These neurons, that form part of the phylogenetically modern networks that regulate drinking behaviour in mammals, are an important target for ANG II. Calcitonin is thought to have an excitatory effect on the subfornical neurons. In addition, it is known that calcitonin surges, as occur during food intake, may be involved in the initiation of the drinking behaviour associated with food intake. Fish lack the analogous network (with the subfornical neurons as key target) for drinking as seen in mammals; drinking in fish is a result of activation of more caudal networks in the fish brain that control swallowing (see Chapter 5, Takei, 2002; this volume). It would be interesting to see whether calcitonin or its gene-related peptide affects such caudal brain targets. An antagonistic relationship with the inhibitory heart peptides (Galli and Phillips, 1996; Takei, 2001) seems an attractive hypothesis to test. Calcitionin gene-related peptide influences the electrophysiology of tilapia intestine (Kiliaan *et al.*, 1993) and thus the hormone could contribute to control of the physiological consequences of drinking.

4. Drinking hormones and antigens

The experiments discussed above on the effects of PTHrP in sea bream are an example where the fish drinks a hormone solution and where a small proportion (1–2%) of the hormone is indeed recovered in the plasma, at least when such experiments are carried out with sufficiently large fish (Guerreiro *et al.*, 2001). The extrapolation of the hormone uptake conditions from adults to larvae is of course an exercise full of pitfalls, but proceeding from the PTHrP-bioactivity observed in the larvae it seems reasonable to assume that significant amounts of bioactive and immunoreactive (plasma levels after PTHrP exposure in the water were assessed by radioimmunoassay) species of PTHrP enter the circulation of the fish/larvae and exert their effect. Furthermore, we cannot exclude that waterborne PTHrP exerted effects via direct activation of PTHrP receptors in apical membranes in the branchial epithelium exposed to the ambient water (as would normally do locally secreted PTHrP?), and by doing so may exert calcemic effects directly on the gills. Messenger for PTHrP and the expressed protein are found in the chloride cells of fish (Ingleton and Danks, 1996; P. Ingleton and G Flik, personal communication); maybe these cells produce this protein as an autocrine substance? At the moment no further information is available on this topic in fish, but the possibility of apical membrane receptors for protein signals—that were long thought to exclusively activate receptors in the basolateral membrane compartment—can no longer be ignored: in rat kidney the PTH/PTHrP receptor is expressed in both the apical and the basolateral plasma membrane domains of the tubular cells and thus control via apical pathways is possible (Traebert *et al.*, 2000).

There is a considerable literature demonstrating that proteins (antigens, hormones) are taken up by fish via intestinal routes (see e.g. McLean *et al.*, 1999; Joosten *et al.*, 1997) and thus drinking of waterborne antigen/protein is a mode of uptake. Important to mention is that addition of a protein to food and processing of food and protein in the stomach may drastically decrease bioactivity as was, for instance, elegantly demonstrated for growth hormone in Japanese eel (Duan and Hirano, 1991). Still (homologous), growth hormone preparations administered orally or by immersion (plus the stress-induced drinking?) improved growth performance in salmonids (Moriyama and Kawauchi, 1990), flounder (Jeh *et al.*, 1998) and gilthead sea bream (Ben Atia *et al.*, 1999). Protecting growth hormone from gastric breakdown and loss of bioactivity was realized by including it in a polymer matrix (rainbow trout; Moriyama *et al.*, 1993) or in sodium alginate- or gelatin-coated yeast containing the inclusion bodies with the expressed recombinant growth hormone (striped mullet; Tsai *et al.*, 1994). There are three relevant additional considerations here. First, some species of fish (like carp) lack a stomach and the acid proteolytic start of digestion and in such fish loss of bioactivity may be less of an item. Second, solutions drunk do not necessarily enter the stomach and as a result bioactivity may escape from gastric breakdown. Third, fish appear to take up proteins and hormones with bioactivity preserved (Joosten *et al.*, 1995); passage of protein hormones into the bloodstream is more efficient in more distal segments (colon) as shown directly by anal intubation studies (Duan and Hirano, 1991; Schep *et al.*, 1998). Furthermore, the organization of the gastrointestinal tract of the fish under study should be taken into account as the variation in intestinal tracts (herbivores, omnivores, carnivores etc.) among the estimated 35 000 species of fish (T. Iwamoto, California Academy of Sciences, personal communication) is phenomenal.

4.1 *Drinking of IGF-I by tilapia larvae*

When newly hatched tilapia (9–10 mm length) born in 60% SW are exposed to homologous recombinant insulin-like growth factor I (IGF-I) in the water, they may be assumed to drink the growth factor. IGF-I is the local mediator of growth hormone effects, also in the gills that express the genes for IGF-I (Reinecke *et al.*, 1997) and it was predicted that IGF-I as a mediator of growth factor (GH) effects (e.g. SW adaptation; Madsen and Bern, 1993) should target the branchial epithelium, for instance by enhancing epithelial turn-over and thus apoptosis.

Indeed, waterborne IGF-I dose-dependently enhances apoptosis in branchial epithelium of the baby tilapia (*Figure 4*). In this way, using the sensitive read-out provided by a commercial apoptosis assay, it is established that in these very early

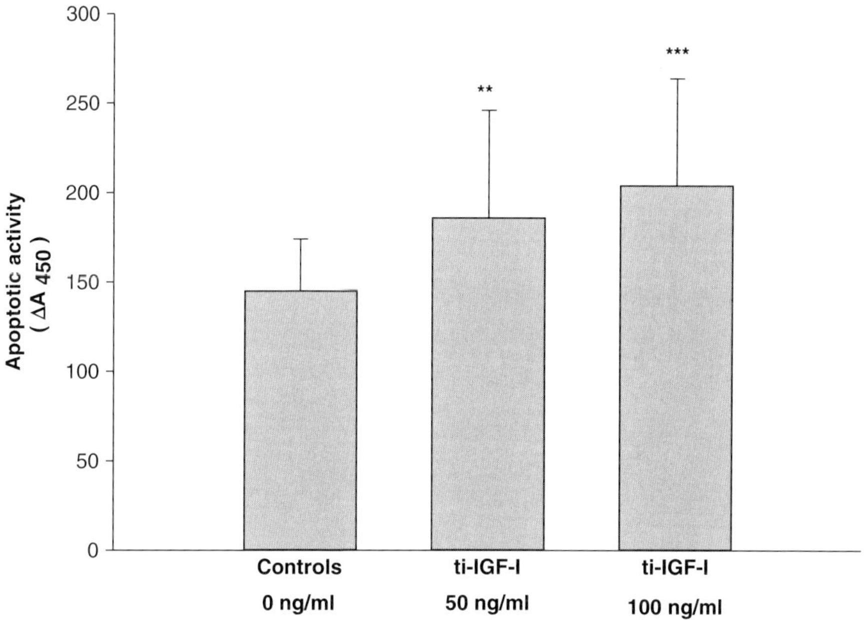

Figure 4. *Effects of waterborne IGF-I on gill apoptosis in larval tilapia. Baby tilapia (9–10 mm body length) were exposed to BrdU (30 mg l^{-1} filtered 60% SW) for 24 h. 5-Bromodeoxyuridine is lipophilic and easily taken up from the water and incorporated into the DNA of the rapidly growing/developing fish and subsequently used to recognize BrdU-labelled DNA-fragments (resulting from apoptotic events) with a highly sensitive and specific anti-BrdU Fab-fragment peroxidase conjugate to allow for detection through production of chromogen; specificity is further enhanced by immobilizing the DNA-fragments to the microtitre plate well by use of a highly specific anti-DNA antibody (Cellular DNA fragmentation Elisa, Roche Cat No. 1 585 045). After* in vivo *BrdU-labelling, the larvae were exposed to recombinant tilapia IGF-I (0, 50 or 100 ng ml^{-1}) for 24 h. The fish were quickly euthanized in phenoxyethanol (0.5%) and the gills excised using a binocular microscope. All eight gill arches per fish were homogenized in the lysis buffer provided with the kit and the assay carried out strictly following the instructions of the manufacturer. Data are given for six individuals per treatment. Apoptosis incidence is expressed in ΔA_{450}/8 gill arches. In a separate experiment using 0, 1, 2, 4 and 8 gill arches of similarly sized fish, it was assessed that a significant apoptosis signal is obtained with this kit from at least two gill arches of fish of this size ($\Delta A_{450} = 10.375n + 114.875$; $r^2 = 0.96625$, P<0.001).*

stages apoptosis is an important mode in IGF-I mediated regulation of the branchial epithelium, that may be crucial in the adaptation of the larvae to, for example, varying salinity.

4.2 *Enhancing drinking in a freshwater fish*

Uptake of vaccine is enhanced by a high-salt shock treatment prior to the vaccine bath immersion (Ototake *et al.*, 1996), but the precise mechanism for enhancement of vaccine uptake under such conditions has not been reported. Drinking rate was measured in 2- and 7-days post-hatching carp larvae kept in Nijmegen tapwater that were netted (mild stressor) or NaCl-shocked (3.5–5.3% in tapwater; severe stressor) for 2 min (*Figure 5*). An essentially adult-type stress response is already operational in these early developmental stages of carp, as indicated by cortisol responses to handling and netting (Flik *et al.*, 2002). If the stressor induced drinking, then controlled application of a stressor may have practical use in aquaculture settings.

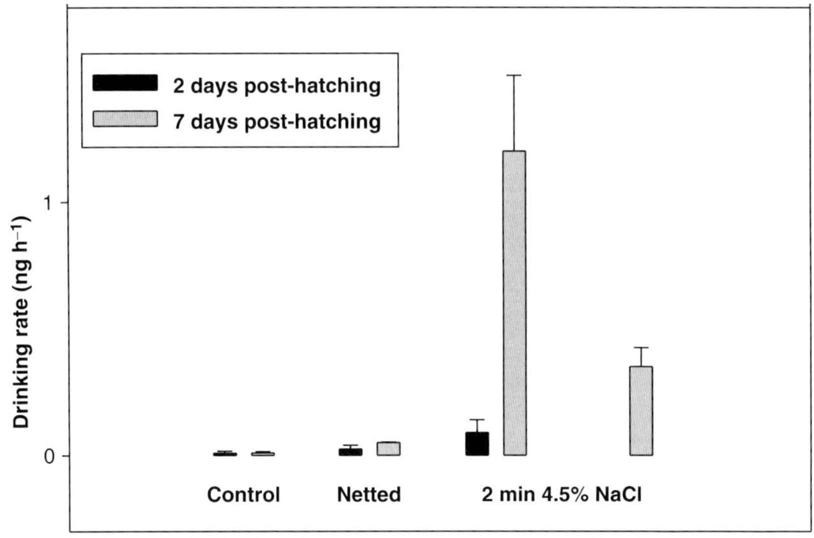

Figure 5. *Drinking in carp (*Cyprinus carpio*) larvae 2 and 7 days post-hatching. Drinking was estimated as described for sea bass in* Figure 1. *Control fish were kept in 1-l containers. To evaluate the effects of a mild stress on drinking, larvae were transferred twice by netting with a 2 min interval immediately before the start of the tracer experiment. To evaluate the stress of exposure to a severe hyper-ionic medium larvae were transferred by netting to 4.5% NaCl for two min and then again transferred to normal water spiked with tracer to estimate drinking. By doing so the 'netted' group could serve as control for the latter. Exposure to high NaCl medium is practised in aquaculture to enhance vaccine uptake. Carp, stenohaline FW fish, drink very little if at all (controls: 9 ± 8 and 10 ± 7 nl h^{-1} per fish; just significantly different from 0,* $P<0.05$*); a mild stressor induces very significant drinking ('netted': 25 and 50 nl h^{-1} per fish); 2 min exposure to a 5.3% NaCl solution, a more severe stressor than netting, increases drinking to a much higher degree, but such treatment causes damage of the integument. Damage is indicated by the difference in 'drinking rate' assessed on the basis of whole body tracer content or intestinal tract tracer content, which gives a lower and realistic value (data only for larvae 7 days post-hatching, second and third bars in last group).*

Drinking rate was assessed on the basis of whole body counts for the water marker ^{51}Cr-EDTA (left six bars in *Figure 5*), and netting induced significant drinking as did the NaCl shock. We got somewhat suspicious though from the high drinking rates in the 7-days post-hatching group and inspected the integument for intactness. Exposure to a high NaCl solution resulted in tissue damage and rather severe apoptosis (M.O. Huising, personal observations) as seen by electron microscopy (TEM and SEM); accordingly, we observed a significant uptake of antigen (BSA-FITC) directly from the water directly into the branchial epithelium (*Figure 6*). Drinking is therefore easily overestimated using this method (based on whole body rather than gut tracer content) after such a treatment.

Indeed when fish were snap-frozen after exposure to high NaCl and ^{51}Cr-EDTA, and when the intestinal tract compartment was excised, a much lower ^{51}Cr-EDTA

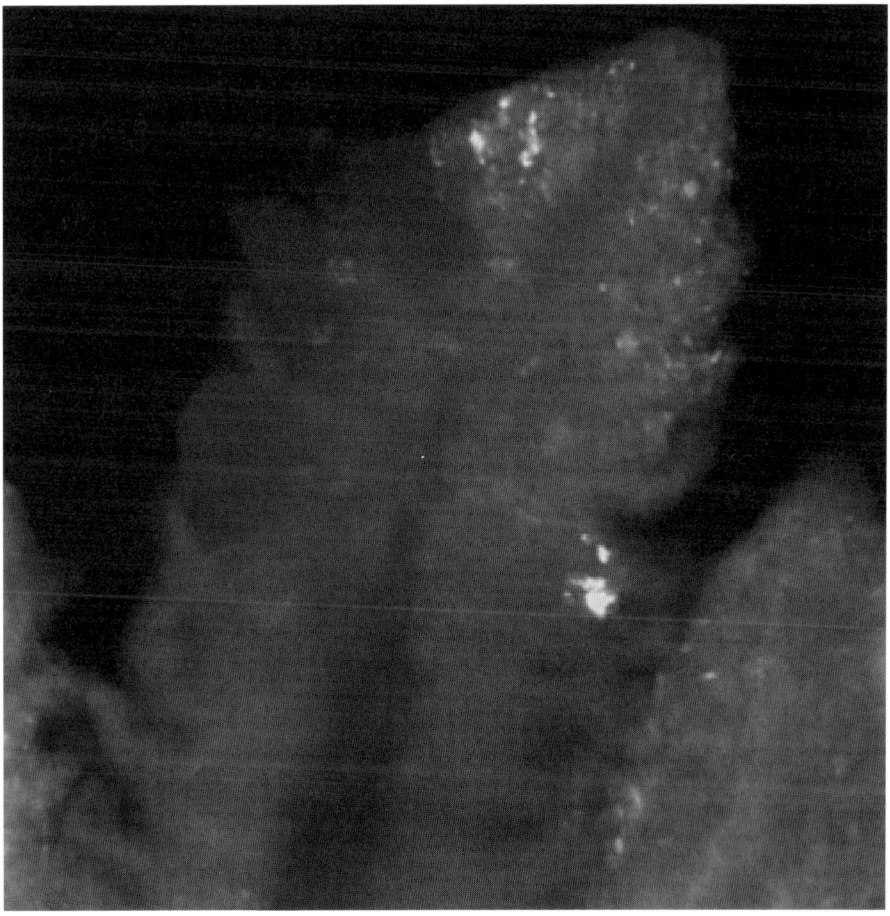

Figure 6. *BSA-FITC fluorescence in gill filament of carp (*Cyprinus carpio *L.) exposed for 2 min to 5.3% NaCl in FW followed by exposure to BSA-FITC (2 mg ml^{-1}) for 10 min; fish were rinsed and sampled 10 min after the exposure to antigen was stopped. Fluorescence has penetrated the branchial epithelium and appears to be confined to cells. Electron microscopy (TEM and SEM) revealed that the NaCl-shock caused significant damage of the epithelium. In controls no BSA-FITC enters the branchial epithelium (data not shown).*

content was found in association with the gut compartment than with the intact larva (*Figure 5*, right bar) and this confirms that significant amounts of tracer entered the larvae via other routes than the gut when the larva is NaCl-shocked. The experiment shows that drinking can be induced by mild and severe stressors. The existence of two different pathways that may differ in their specific immune system components should be considered: the tight epithelium of the gills (Santos *et al.*, 2001) and the more leaky epithelium of the gut (Azad *et al.*, 2000). In fish exposed to waterborne antigen (Olafsen, 2001) both epithelia should be studied and evaluated in fish immunization and vaccination programs.

Acknowledgements

The experiment mentioned under **4.1** was carried out in the laboratory of Prof. Dr H.A. Bern during a sabbatical stay of G.F. in Berkeley. The hospitality of the Berkeley crew and skilful help of Isabel Hsu are most heartily acknowledged.

References

Azad, I.S., Shankar, K.M., Mohan, C.V. and Kalita, B. (2000) Uptake and processing of biofilm and free-cell vaccines of *Aeromonas hydrophila* in Indian major carps and common carp following oral vaccination – antigen localization by a monoclonal antibody. *Diseases Aquat. Org.* **43:** 103–108.

Azria, M., Copp, D.H. and Zanelli, J.M. (1995) 25 years of salmon calcitonin – from synthesis to therapeutic use. *Calcif. Tissue Int.* **57:** 405–408.

Balment, R.J., Warne, J.M., Tierney, M. and Hazon, N. (1993) Arginine vasotocin and fish osmoregulation. *Fish Physiol. Biochem.* **11:** 189–194.

Barnabé, G., Boulineau-Coatanea, F. and Rene, F. (1976) Chronologie de la morphogenèse chez le loup ou bar *Dicentrarchus labrax* (L.) (Pisces, Serranidae) obtenu par reproduction artificielle. *Aquaculture* **8:** 351–363.

Ben-Atia, I., Fine, M., Tandler, A., Funkenstein, B., Maurice, S., Cavari, B. and Gertler, A. (1999) Preparation of recombinant gilthead sea bream (*Sparus aurata*) growth hormone and its use for stimulation of larvae growth by oral administration. *Gen. Comp. Endocrinol.* **113:** 155–164.

Bern, H.A. (1975) Prolactin and osmoregulation. *Am. Zool.* **15:** 937–948.

Brown, J.A. and Tytler, P (1993) Hypoosmoregulation of larvae of the turbot, *Scophthalmus maximus*. Drinking and gut function in relation to environmental salinity. *Fish Physiol. Biochem.* **10:** 475–483.

Butler, D.G. and Alia Cadinouche, M.Z. (1995) Fractional reabsorpition of calcium, magnesium and phosphate in the kidneys of freshwater North American eels (*Anguilla rostrata* LeSueur) following removal of the corpuscles of Stannius. *J. Comp. Phys.* **165B:** 348–358.

Butler, D.G. and Oudit, G.Y. (1995) Corpuscles of Stannius and blood flow regulation in freshwater North American eels, *Anguilla rostrata* LeSueur. *J. Endocrinol.* **145:** 181–194.

Chester Jones, I., Henderson, I.W., Chan, D.K.O., Rankin, J.C., Mosley, E., Brown, J.J., Lever, A.F., Robertson, J.I.S. and Tree, M. (1966) Pressor activity in extracts of the corpuscles of Stannius from the European eel *Anguilla anguilla* L. *J. Endocrinol.* **34:** 339–408.

Copp, D.H., Ma, S.W.Y., Qui, M.C., Sim, M. and Szeto, K. (1985) Role of the pituitary and the corpuscles of Stannius and teleocalcin in calcium regulation in teleosts. In: *Current Trends in Comparative Endocrinology* (eds B. Lofts and W.N. Holmes). Hong Kong University Press, Hong Kong, pp. 819–821.

Duan, C. and Hirano T. (1991) Plasma kinetics of growth hormone in the Japanese eel, *Anguilla japonica*. *Aquaculture* **95:** 179–188.

Ellis, A.E. (1998) Meeting the requirements for delayed release of oral vaccines for fish. *J. Appl. Ichtyol.* **14:** 149–152.

Fenwick, J.C., Flik, G. and Verbost, P.M. (1995) A passive immunization technique against the teleost hypocalcemic hormone stanniocalcin provides evidence for the cholinergic control of stanniocalcin release and the conserved nature of the molecule. *Gen. Comp. Endocrinol.* **98:** 202–210.

Flik, G. (1990) Hypocalcin physiology. In: *Progress in Comparative Endocrinology*. Wiley-Liss, pp. 578–585.

Flik, G., Wendelaar Bonga, S.E. and Fenwick, J.C. (1984) Ca^{2+}-dependent phosphatase and Ca^{2+}-dependent ATPase activities in plasma membranes of eel gill epithelium – III. Stimulation of branchial high-affinity Ca^{2+}-ATPase activity during prolactin-induced hypercalcemia in American eels. *Comp. Biochem. Physiol.* **79B:** 521–524.

Flik, G., Fenwick, J.C., Kolar, Z., Mayer-Gostan, N. and Wendelaar Bonga, S.E. (1986) Effects of ovine prolactin on calcium uptake and distribution in *Oreochromis mossambicus*. *Am. J. Physiol.* **250:** R161–R166.

Flik, G., Fenwick, J.C. and Wendelaar Bonga, S.E. (1989) Calcitropic actions of prolactin in freshwater North American eel (*Anguilla rostrata* LeSueur). *Am. J. Physiol.* **257:** R74–R79.

Flik, G., Rentier-Delrue, F. and Wendelaar Bonga, S.E. (1994) Calcitropic effects of recombinant prolactins in *Oreochromis mossambicus*. *Am. J. Physiol.* **266:** R1302–R1308.

Flik, G., Stouthart, X.J.H., Spanings, F.A.T., Lock, R.A.C., Fenwick, J.C. and Wendelaar Bonga, S.E. (2002) Stress response to waterborne Cu during early life stages of carp, *Cyprinus carpio*. *Aqua. Tox.* **56:** 167–176.

Fuentes, J. and Eddy, F.B. (1997) Drinking in marine, euryhaline and freshwater teleost fish. In: *Ionic Regulation in Animals* (eds N. Hazon, F.B. Eddy. and G. Flik). Springer, Heidelberg/New York, pp. 135–149.

Galli, S.M. and Phillips, M.I. (1996) Interactions of angiotensin II and atrial natriuretic peptide in the brain: fish to rodent. *Proc. Soc. Exp. Biol. & Med.* **213:** 128–137.

Guerreiro, P.M., Fuentes, J., Power, D.M., Ingleton, P.M., Flik, G. and Canario, A.V.M. (2001) Parathyroid hormone-related protein: a calcium regulatory factor in sea bream (*Sparus aurata* L.) larvae. *Am. J. Phys.* **281:** R855–R860.

Hanssen, R.G.J.M., Lafeber, F.P.J.G., Flik, G. and Wendelaar Bonga, S.E. (1989) Ionic and total calcium levels in the blood of the European eel (*Anguilla anguilla*): effects of stanniectomy and hypocalcin replacement therapy. *J. Exp. Biol.* **141:** 177–186.

Heijden, A.J.H. van der, Meij, J.C.A. van der, Flik, G. and Wendelaar Bonga, S.E. (1999a) Ultrastructure and distribution dynamics of chloride cells in tilapia larvae in fresh water and sea water. *Cell Tissue Res.* **297:** 119–130.

Heijden, A.J.H. van der, Verbost, P.M., Bijvelds, M.J.C., Atsma, W., Wendelaar Bonga, S.E. and Flik, G. (1999b) Effects of seawater and stanniectomy on branchial Ca^{2+} handling and drinking rate in eel (*Anguilla anguilla* L.). *J. Exp. Biol.* **202:** 2505–2511.

Ingleton, P. and Danks, J. (1996) Distribution and functions of parathyroid hormone-related protein in vertebrate cells. *Int. Rev. Cytol.* **166:** 231–280.

Jeh, H.S., Kim, C.H., Lee, H.K. and Han, K. (1998) Recombinant flounder growth hormone from *Escherichia coli*: overexpression, efficient recovery and growth-promoting effects on juvenile flounder by oral administration. *J. Biotechn.* **60:** 183–193.

Joosten, P.H.M., Avilestrigueros, M., Sorgeloos, P. and Rombout, J.H.W.M. (1995) Oral vaccination of juvenile carp (*Cyprinus carpio*) and gilthead sea bream (*Sparus aurata*) with bioencapsulated vibrio anguillarum bacterin. *Fish Shellfish Immunol.* **5:** 289–299.

Joosten, P.H.M., Engelsma, M.Y., Zee, M.D. van der and Rombout, J.H.W.M. (1997) Induction of oral tolerance in carp (*Cyprinus carpio* L.) after feeding protein antigens. *Vet. Immunol. Immunopath.* **60:** 187–196.

Kiliaan, A.J., Holmgren, S., Jonsson, A.C., Dekker, K. and Groot, J.A. (1993) Neuropeptides in the intestine of 2 teleost species (*Oreochromis mossambicus*, *Carassius auratus*)—localization and electrophysiological effects on the epithelium. *Cell Tissue Res.* **271:** 123–134.

Madsen, S.S. and Bern, H.A. (1993) In-vitro effects of insulin-like growth factor-I on gill Na^+,K^+-ATPase in Coho salmon, *Oncorhynchus kisutch*. *J. Endocrinol.* **138:** 23–30.

McLean, E., Ronsholt, B., Sten, C. and Najamuddin, A. (1999) Gastrointestinal delivery of peptide and protein drugs to aquacultured teleosts. *Aquaculture* **177:** 231–247.

Moriyama, S. and Kawauchi, H. (1990) Growth stimulation of juvenile salmonids by immersion in recombinant salmon growth hormone. *Nippon Suisan Gakkaishi* **56:** 31–34.

Moriyama, S., Yamamoto, H., Sugimoto, S., Abe, T., Hirano, T. and Kawauchi, H. (1993) Oral administration of recombinant salmon growth hormone to rainbow trout, *Oncorhynchus mykiss*. *Aquaculture* **112:** 99–106.

Musch, M.W., Orellana, S.A., Kimberg, L.S., Fiel, M., Halm, D.R., Kransky, E.J. and Fritzzell, R.A. (1982) Na-K-Cl cotransport in the intestine of a marine teleost. *Nature* **300:** 351–354.

Olafsen, J.A. (2001) Interactions between fish larvae and bacteria in marine aquaculture. *Aquaculture* **200:** 223–247.

Ototake, M., Iwama, G.K. and Nakanishi, T. (1996) The uptake of bovine serum albumin by the skin of bath-immunised rainbow trout *Oncorhynchus mykiss*. *Fish Shellfish Immunol.* **6:** 321–333.

Pang, P.K.T., Griffith, R.W., Maetz, J. and Pic, P. (1980) Calcium uptake in fishes. In: *Epithelial Transports in the Lower Vertebrates* (ed. B. Lahlou). Cambridge University Press, Cambridge, pp. 121–131.

Pierson, P.M., Guibbolini, M.E., Mayer Gostan, N. and Lahlou, B. (1995) Elisa measurements of vasotocin and isotocin in plasma and pituitary of the rainbow trout—effect of salinity. *Peptides* **16:** 859–865.

Potts, W.T.W. and Eddy, F.B. (1973) The permeability to water in the eggs of certain marine teleosts. *J. Comp. Physiol.* **87:** 305–315.

Reinecke, M., Schmid, A., Ermatinger, R. and Loffing Cueni, D. (1997) Insulin-like growth factor I in the teleost *Oreochromis mossambicus*, the tilapia: gene sequence, tissue expression and cellular localization. *Endocrinol.* **138:** 3613–3619.

Reitan, K.I., Natvik, C.M. and Vadstein, O. (1998) Drinking rate, uptake of bacteria and microalgae in turbot larvae. *J. Fish Biol.* **53:** 1145–1154.

Ruohonen, K., Grove, D.J. and McIlroy, J.T. (1997) The amount of food ingested in a single meal by rainbow trout offered chopped herring, dry and wet diets. *J. Fish. Biol.* **51:** 93–105.

Santos, N.M.S. dos, Taverne-Thiele, J.J., Barnes, A.C., Muiswinkel, W.B. van, Ellis, A.E. and Rombout, J.H.W.M. (2001) The gill is a major organ for antibody secreting cell production following direct immersion of sea bass (*Dicentrarchus labrax*, L.) in a *Photobacterium damselae* ssp *piscicida* bacterin: an ontogenetic study. *Fish Shellfish Immunol.* **11:** 65–74.

Schep, L.J., Tucker, I.G., Young, G., Ledger, R. and Butt, A.G. (1998) Permeability of the salmon (*Oncorhynchus tshawytscha*) posterior intestine in vivo to two hydrophilic markers. *J. Comp. Physiol. B* **168:** 562–568.

Schmid, H.A., Rauch, M. and Koch, J. (1998) Effect of calcitonin on the activity of angiotensin II-responsive neurons in the rat subfornical organ. *Am. J. Phys.* **43:** R1646–R1652.

Schoenmakers, Th.J.M., Verbost, P.M., Flik, G. and Wendelaar Bonga, S.E. (1993) Transcellular intestinal calcium transport in freshwater and seawater fish and its dependence on sodium/calcium exchange. *J. Exp. Biol.* **176:** 195–206.

Takei, Y. (2000) Comparative physiology of body fluid regulation in vertebrates with special reference to thirst regulation. *Jap. J. Physiol.* **50:** 171–186.

Takei, Y. (2001) Does the natriuretic peptide system exist throughout the animal and plant kingdom? *Comp. Biochem. Physiol.* **129:** 559–573.

Tierney, M.L., Luke, G., Cramb, G. and Hazon, N. (1995) The role of the renin angiotensin system in the control of blood pressure and drinking in the European eel, *Anguilla anguilla. Gen. Comp. Endocrinol.* **100:** 39–48.

Traebert, M., Volkl, H., Biber, J., Murer, H. and Kaissling, B. (2000) Luminal and contraluminal action of 1–34 and 3–34 PTH peptides on renal type IIa Na-P_i cotransporter. *Am. J. Phys.* **278:** F792–F798.

Tsai, H.J., Kuo, J.C., Lou, S.W. and Kuo, T.T. (1994) Growth enhancement of juvenile striped mullet by feeding recombinant yeasts containing fish growth-hormone. *Prog. Fish Culturist* **56:** 7–12.

Tsuchida, T. and Takei, Y. (1999) A potent dipsogenic action of homologous angiotensin II infused at physiological doses in eels. *Zool. Sci.* **16:** 479–483.

Tytler, P. and Blaxter, J.H.S. (1988a) The effects of external salinity on the drinking rates of the larvae of herring, plaice and cod. *J. Exp. Biol.* **138:** 1–15.

Tytler, P. and Blaxter, J.H.S. (1988b) Drinking in yolk-sac stage larvae of the halibut, *Hipoglossus hipoglossus* (L.). *J. Fish Biol.* **32:** 493–494.

Tytler, P. and Ireland, J. (1994) Drinking and water-absorption by the larvae of herring (*Clupea harengus*) and turbot (*Scophthalmus maximus*). *J. Fish. Biol.* **44:** 103–116.

Varsamos, S., Diaz, J.P., Charmantier, G., Blasco, C., Connes, R. and Flik, G. (2002) Location and morphology of chloride cells during the post-embryonic development of the European sea bass, *Dicentrarchus labrax. Anat. Embryol.* (in press).

Verbost, P.M., Flik, G., Fenwick, J.C., Greco, A.M., Pang, P.K.T. and Wendelaar Bonga, S.E. (1993) Branchial calcium uptake: possible mechanisms of control by stanniocalcin. *Fish Physiol. Biochem.* **11:** 205–215.

Vonck, A.P.M.A., Wendelaar Bonga, S.E. and Flik G. (1998) Sodium and calcium balance in Mozambique tilapia, *Oreochromis mossambicus*, raised at different salinities. *Comp. Biochem. Physiol.* **119:** 441–449.

Wendelaar Bonga, S.E. (1980) Effects of synthetic salmon calcitonin and low ambient calcium on plasma calcium, ultimobranchial cells, Stannius bodies and prolactin cells in the teleost *Gasterosteus aculeatus. Gen. Comp. Endocrinol.* **40:** 99–108.

Wendelaar Bonga, S.E. and Lammers, P.I. (1982) Effects of calcitonin on ultrastructure and mineral content of bone and scales of the cichlid teleost *Sarotherodon mossambicus. Gen. Comp. Endocrinol.* **48:** 60–70.

Wendelaar Bonga, S.E., Flik, G., Löwik, C.W.G.M. and Van Eijs, G.J.J.M. (1985) Environmental control of prolactin synthesis in the teleost fish *Oreochromis* (formerly *Sarotherodon*) *mossambicus. Gen. Comp. Endocrinol.* **57:** 352–359.

Yamamoto, S., Morimoto, I., Zeki, K., Ueta, Y., Yamashita, H., Kannan, H. and Eto, S. (1998) Centrally administered parathyroid hormone (PTH)-related protein(1–34) but not PTH(1–34) stimulates arginine-vasopressin secretion and its messenger ribonucleic acid expression in supraoptic nucleus of the conscious rat. *Endocrinology* **139:** 383–388.

Zadunaiski, J.A. (1997) Gill chloride cells activation by plasma osmolarity. In: *Ionic Regulation in Animals* (eds N. Hazon, F.B. Eddy and G. Flik). Springer, Heidelberg/New York, pp. 87–103.

4

Water transport and aquaporins in the European eel (*Anguilla anguilla*)

J-H. Lignot, C.P. Cutler, N. Hazon and G. Cramb

1. Introduction

A fundamental problem for euryhaline teleosts is the maintenance of a body fluid composition different from that of their immediate environment. In order to osmoregulate, these fish employ the functions of a number of tissues including the gill, oesophagus, intestine, kidney and urinary bladder. It is now well established that freshwater (FW) teleosts drink little and produce large volumes of urine to compensate for osmotic water uptake through the gills. In contrast, marine or sea-water (SW) teleosts ingest relatively large amounts of SW and absorb most of this water along with monovalent ions across the intestine. Together with active excretion of salts by the gills and other organs, this osmoregulatory strategy compensates for osmotic water loss and dehydration when in SW.

To date, substantial progress has been made in our understanding of the molecular basis of ion homeostasis in teleost fish and a number of ion and solute transporters have been identified. In this respect, the eel has proved to be a valuable model for osmoregulatory studies as this species can survive acute transfers between either SW or FW (Tesch and Greenwood, 1977). As with most euryhaline teleosts, the eel is therefore able to modify the ion and water permeability of its osmoregulatory epithelial surfaces following environmental salinity change with the gills and the digestive tract being the major sites of water exchange between the internal milieu and the external environment. However, despite its vital significance, there is currently a lack of molecular information on water transport mechanisms associated with these osmoregulatory tissues. New information on the type, characteristics and regulation of water transporters are central to the understanding of this fundamental process.

Osmoregulation and Drinking in Vertebrates, edited by N. Hazon & G. Flik.
© 2002 BIOS Scientific Publishers Ltd, Oxford

2. Water transport in fish

2.1 *Water fluxes across the digestive tract*

Of the limited amount of information available concerning water flux pathways across the digestive tract, most data has been gathered using the eel. When eels are transferred to the marine environment they immediately drink SW which enters the oesophagus at a slow but steady rate (Parmelee and Renfro, 1983). Here salts are absorbed by both secondary active and passive processes, and as water permeability of the oesophagus is extremely low only small net effluxes of water occur due to osmosis. This leads to an overall dilution of the oesophageal fluid concentration to around one-half to one-third of the initial osmolality (Hirano and Mayer-Gostan, 1976; Nagashima and Ando, 1993; Parmelee and Renfro, 1983). The oesophageal desalination appears therefore to be an essential process. In the stomach, very little further ion uptake or water secretion occurs (Hirano and Mayer-Gostan, 1976). When the luminal fluid reaches the intestine, water is initially absorbed against the osmotic gradient together with salts. Absorption occurs until the Na and Cl concentrations are reduced to around 30–90 meq l^{-1}, and most of the luminal fluid volume has been taken up into body fluids (Hirano and Mayer-Gostan, 1976; Parmelee and Renfro, 1983; Skadhauge, 1969). Water transport in the intestine occurs by two processes: either 1) secondary-active water transport (which can transport water against the osmotic gradient) thought to occur through a trans-/para-cellular shunt pathway and/or 2) passive water fluxes which follow the osmotic gradients and which are entirely trans-cellularly based (Skadhauge, 1969). The secondary-active water transport is tightly coupled to the transport of chloride and sodium and/or potassium ions (Skadhauge, 1969, 1974; Ando, 1975, 1980, 1981, 1985) and has been described as solute-coupled water flux.

The capacity of this solute-linked water flux increases with acclimation to increasing salinities (Skadhauge, 1969, 1974) with this component increasing 3–3.4-fold following FW to SW acclimation (Utida *et al.*, 1972; Ando, 1975). The passive osmotic water permeability also increases by 2–6-fold following SW acclimation (Skadhauge, 1969; Ando, 1975). Finally, the magnitude of water fluxes across various parts of the intestine have also been determined, with highest levels occurring in the mid region followed in descending order by the posterior, anterior intestine and rectum (Ando and Kobayashi, 1978; Ando, 1981).

2.2 *Water fluxes across the gills*

The gills are the major site of water exchange between the internal milieu and the external aquatic environment. In general, osmotic water permeability is higher in FW- than in SW-acclimated fish, indeed in the FW-acclimated European eel, *Anguilla anguilla*, osmotic water permeability is six times greater than in SW (Isaia, 1984). In FW, the branchial epithelium is responsible for over 90% of the total body water influx (Motais *et al.*, 1969; Isaia, 1984). Furthermore, diffusional water fluxes across the gill occur primarily across the secondary lamellae due in part to the large surface area.

The principal osmotic water flux pathway across the gill has been suggested to operate through the chloride cells (mitochondria-rich, ion-transporting cells) located primarily within the branchial primary filaments in SW-acclimated eels and along the base of the secondary filaments in FW-acclimated eels. This process has been associated with 'bulk flow' water and ion transporting mechanism via specialized membrane structures of the chloride cells; the baso-lateral tubular network, which results in water efflux through 'leaky' tight junctions into the external environment (Isaia, 1984). While this may explain a portion of gill water permeability in marine fish, fluxes through the tubular network of chloride cells in FW teleosts are likely to be lower, as there are fewer chloride cells with more extensive tight junctions (Potts, 1984). This suggests that osmotic water fluxes (particularly in FW fish) must also occur across other surfaces of the branchial epithelium. Included in the different membrane compartments which have been hypothesized to be involved in the transfer of solutes across the gills, is the tubulo-vesicular system situated between the apical membrane of the chloride cells and the extensive intracellular tubular reticulum (Sardet *et al.*, 1979). This space could represent a transient communication channel between the internal and external milieu (Sardet *et al.*, 1979). It has also been suggested that other surface epithelial cells such as the 'respiratory cells' located along the secondary lamellae of the gills, may be involved in osmotic water fluxes (Rankin and Bolis, 1984). While it is not clear why osmotic water permeability should be so much greater in FW- compared to SW-acclimated eels, it does suggest that a regulated pathway for water movement in the gill may exist, the components of which would be expected to be more abundant in FW than SW fish.

3. Facilitated water transporters (aquaporins)

Aquaporins are members of the family of Major Intrinsic Proteins (MIPs). Their central role in the regulation of water transport across most epithelia is now well established. Indeed they have been implicated in numerous physiological processes as well as the pathophysiology of a wide range of clinical disorders (Preston *et al.*, 1994; King and Agre, 1996; Sasaki *et al.*, 1998; Borgnia *et al.*, 1999; Verkman and Mitra, 2000).

To date, many members of the aquaporin (AQP) family have been discovered in mammals, amphibians as well as plants, yeast and bacteria (Park and Saier, 1996). Structurally there are typically six transmembrane domains, two halves of a homologous repeat plus two conserved 'NPA' boxes (amino acid sequences of asparagine–proline–alanine). They function as membrane channels or pores and are either solely concerned with the transport of water (AQP1–2, AQP4–6: orthodox aquaporins or AQP cluster) or with water and other solutes such as urea and/or glycerol along with other small solutes (AQP0, AQP3, AQP7 and AQP9). This second group (with the exception of AQP0) belong to the aquaglyceroporins or glycerol facilitator-like (GLP cluster) proteins (Raina *et al.*, 1995; Echevarria *et al.*, 1996; Ishibashi *et al.*, 1997a; Ishibashi *et al.*, 1998; Yang and Verkman, 1997; Borgnia *et al.*, 1999; Heymann and Engel, 1999). AQP8 may also be permeable to urea but this channel is structurally distinct from other isoforms and has similarity to many plant aquaporins (Ishibashi *et al.*, 1997b; Heymann and Engel, 1999). Two more (AQPX1 and AQPX2), less well-conserved members of the aquaporin family have also recently been identified, but these remain to be fully characterized (Ishibashi *et al.*, 2000). Most aquaporins, with the exception of AQP4 and AQP7, are sensitive to mercuric compounds with inhibition being reversible following application of 2-mercaptoethanol (Echevarria *et al.*, 1994; Ishibashi *et al.*, 1997a). Furthermore, the permeability of glycerol and urea of GLP proteins is three to four orders of magnitude lower than the water permeability

1111371085

suggesting that in these channels a large flux of water accompanies solute permeation (Heymann and Engel, 1999).

4. Aquaporins in fish

4.1 *Evidence for the existence of aquaporins in teleost fish*

Initial studies focused on the cloning and identification of aquaporin homologues in osmoregulatory tissues such as the gill, intestine and kidney, using degenerate RT-PCR. To date, three eel aquaporin homologues have been identified as AQP1, AQP1 dup. and AQP3 (Cutler and Cramb, 2000). The eel AQP1 homologue that shared highest levels of nucleotide and amino acid homology with mammalian AQP1 (Preston and Agre, 1991) has been characterized from the eel intestinal epithelium. By Northern blotting, this AQP1 was shown to have a wide tissue distribution and expression of a 1.4 kb mRNA was found in brain, eye, heart, pancreas, oesophagus, stomach, and intestine, with much lower levels in skeletal muscle, gill and kidney. A minor 3.1 kb mRNA component was also present in some tissues. A duplicate isoform of the AQP1 (AQP1 dup.) has also been characterized from the eel kidney. Finally, a third aquaporin homologue has been identified from the gill and shares highest levels of nucleotide and amino acid homology with mammalian AQP3. Subsequent Northern blotting showed that this AQP3 homologue is expressed as a 2.4 kb mRNA in the eye, oesophagus, intestine and gill with a minor amount of a 7 kb mRNA species also present.

In order to examine the possible physiological role of these water channel homologues during acclimation of fish to different environmental salinities, quantitative studies were carried out. They revealed that the level of mRNA expression for the AQP1 homologue in the intestine increased by 10 to 25-fold 3 weeks after transfer of eels from FW to SW. Messenger RNA abundance was also significantly decreased to 28% of FW values in the kidneys of SW-acclimated fish, although overall levels in the kidney were much lower than the intestine. For the AQP1 dup. isoform initial experiments have indicated that mRNA is only detectable in the oesophagus, intestine and kidney, and that the high level of mRNA expression in the intestine and kidney are dramatically down-regulated when eels are transferred from FW to SW, although there seems to be a considerable amount of heterogeneity in the level of expression between individual animals. For the AQP3 homologue, the major site of mRNA expression was in the gill of FW eels, although the branchial level of expression decreased significantly to 3% of the FW value in SW-acclimated silver eels. In the intestine, the levels of AQP3 mRNA expression were relatively low in both FW and SW fish. Using Western blotting with a specific polyclonal antibody directed against the C-terminal of the eel AQP3, staining indicated the presence of AQP3 proteins mainly within the gills. In a similar fashion to branchial AQP3 mRNA expression, SW acclimation also induced a three-fold decrease in the branchial protein expression. On gill sections of both SW- and FW-acclimated eels, a strong staining indicating the presence of the AQP3 protein was found within the branchial chloride cells. At the subcellular level, AQP3 appeared localized along the baso-lateral tubular system and towards the apical surface of the chloride cells, below the apical pit. Furthermore, in FW-acclimated eels, the localization study clearly demonstrated the presence of the AQP3 water channel expression in cells within the branchial arch and within some of the primary filaments that contain very few chloride cells. This staining predominated in plasma membranes of cells on the serosal side of the epithelium. A reduction in staining in SW-acclimated eels could at least partially explain the three-fold increase in AQP3 protein expression in FW-acclimated compared to SW-acclimated eels. In the digestive tract, AQP3 was only detected in intra-epithelial macrophage-like bodies in the intestine of FW-acclimated eels and in mucous cells of the rectal epithelium of SW-adapted fish.

5. Possible roles for AQP1 and AQP3 in the digestive tract of eels

5.1 *The oesophagus*

The presence of all three aquaporin homologues in the oesophagus is puzzling as this tissue is thought to represent a particularly water-impermeable epithelium (Hirano and Mayer-Gostan, 1976). In FW-acclimated eels the oesophagus is lined by a stratified epithelium consisting of numerous mucous or goblet cells (Yamamoto and Hirano, 1978) and is almost impermeable to both ions and water (Motais *et al.*, 1969; Skadhauge, 1969, 1974; Bentley, 1971). In SW-acclimated eels, the stratified epithelium is replaced by a simple columnar epithelium almost free of mucous cells, is heavily folded and shows a selective ion permeability but a very low water permeability (Laurent and Kirsch, 1975; Yamamoto and Hirano, 1978). Large amounts of sodium and chloride ions therefore move out of the lumen and only a small net amount of osmotic water movement crosses the oesophagus (Skadhauge, 1969; Kirsch *et al.*, 1975; Hirano and Mayer-Gostan, 1976; Kirsch, 1978). However uni-directional fluxes were not measured and larger (nearly equivalent) uni-directional fluxes which result in a very small net flux, could also occur (Parmelee and Renfro, 1983). This suggests that when marine teleosts drink, the oesophagus absorbs both ions and water, but a roughly equivalent amount of water is lost by osmosis to the hyper-osmotic luminal contents. Aquaporins may therefore play a role in either the uptake or loss of water in this tissue.

The difference in the oesophageal mechanisms associated with water transport between SW- and FW-acclimated eels, is illustrated by the location of AQP3 proteins in this organ. In FW-acclimated eels, the strong expression of AQP3 protein only observed in the goblet cells could indicate major water trafficking associated with mucous secretion. A role for maintenance of 'wetness' on the luminal surface of the oesophagus can also be hypothesized, as speculated for AQP5 in corneal squamous epithelial cells in the eye (Funaki *et al.*, 1998). In SW-acclimated eels however, the strong presence of AQP3 protein on the plasma membranes of the basal epithelial cells of the oesophagus indicates the presence of a trans-cellular pathway for water. This localization for AQP3 correlates very well with previous findings obtained for AQP3 from rat oesophagus (Koyama *et al.*, 1999) and suggests water movement from the serosal side to the epithelial cells in order to protect these cells from dehydration, as previously suggested (Matsuzaki *et al.*, 1999; Matsuzaki *et al.*, 2000).

5.2 *The intestine and rectum*

The low levels of AQP3 mRNA and protein expression observed in the intestinal enterocytes of both SW- and FW-acclimated eels, agree with a recent study hypothesizing the absence of water channels in eel brush border membrane vesicles (Alves *et al.*, 1999). AQP3 is therefore unlikely to play a major role in the intestine and especially in the water absorption that occurs following the drinking response in SW-acclimated fish. The strong expression of AQP3 protein in the rectal goblet cells of SW-acclimated eels however, indicates a major role for this protein in water trafficking associated with mucous secretion. A role for maintenance of 'wetness' on the luminal surface of the rectal epithelium can be hypothesized, as for the oesophagus (see previous section).

The strong expression of the AQP1 homologue in the eel intestine suggests that this water channel may play a prominent role in mechanisms associated with the intestinal water absorption, but its exact purpose remains unclear. It seems unlikely that an AQP1 homologue could be directly responsible for secondary active water transport in the intestine. It is however possible that water and salt uptake are mediated by a vesicle shuttling system, which could be used to create a region with high osmolality, using active ion transporters together with aquaporins. It is also possible (and probably more likely) that water can enter together with salts directly through an ion carrier protein such as the Na,K,2Cl co-transporter (Zeuthen, 1995) with water exiting through baso-laterally located aquaporin water channels. As aquaporins are thought to represent passive water channels, it is also possible that AQP 1 may contribute to this passive osmotic permeability.

5.3 *Possible roles for AQP3 in branchial water transport in eels*

The presence of AQP3 along the plasma membranes of basal epithelial cell of FW-acclimated eels (gill arch and primary filament epithelia), is similar to the sub-surface location of this protein in rat urinary bladder epithelia in addition to rat and human skin (Matsuzaki *et al.*, 1999; Matsuzaki *et al.*, 2000). In rats, the role for AQP3 along the plasma membranes of sub-surface epithelial cells is thought to be related to the protection of the cells from dehydration. Such a role in mammals is strengthened by the fact that AQP3 expression in rats is controlled by the osmolality of the extra-cellular fluid (Matsuzaki *et al.*, 1999; Matsuzaki *et al.*, 2000). In the eel, the location of AQP3 along the plasma membranes of basal epithelial cells of FW-acclimated eels also suggests water movement across epithelial cells but its role could be somewhat different. It is hypothesized here that this water movement across the basal cells of the branchial epithelium could occur in order to release water entering apically from the external hypo-osmotic FW environment. This process would prevent cell swelling and eventual lysis since cell volume regulation, in the face of continuous water uptake, could not be easily achieved by the secretion of ions or other osmolytes.

The presence of AQP3 as a water channel within the chloride cells of both FW- and SW-acclimated eels (see *Figure 1*) correlates with physiological and morphological evidence indicating that these cells provide a major route for water as well as ion movement in SW (Isaia, 1984; Ogasawara and Hirano, 1984). The presence of AQP3 along the baso-lateral tubular network of the chloride cells could be related to the osmotic water flux which has been hypothesized to be operating in the tubular system (Isaia, 1984). The sub-apical and possibly apical location (i.e. in the apical pit) of AQP3 also suggests that another regulated pathway for water and/or other solutes may occur within the chloride cells of FW- and SW-acclimated fish. The tubulo-vesicular system situated between the apical plasma membrane of the chloride cells and the tubular reticulum has already been hypothesized as a transient

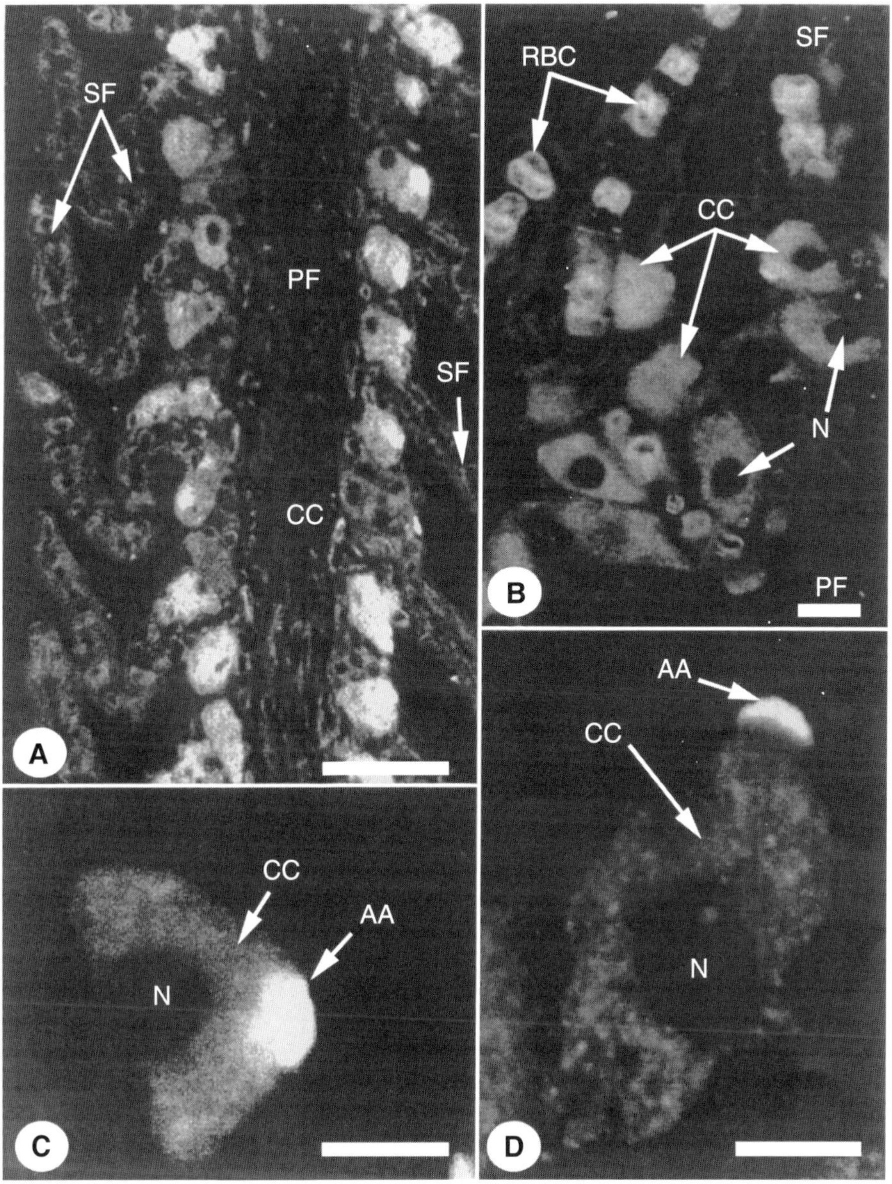

Figure 1. *Immunolocalization of AQP3 in the gill of SW- (A, C) and FW- (B, D) acclimated silver eels. The scale bars correspond to 30 μm in A, 10 μm in B and 5 μm in C and D. AA: apical area; CC: chloride cell; N: nucleus; PF: primary filament; RBC: red blood cell; SF: secondary filament.*

communication channel between the internal and external milieu (Sardet *et al.*, 1979). AQP3 located in this space which is present in both SW- and FW-acclimated fish could therefore participate in water transport via the apical pit. Other possibilities for the binding of the AQP3 antibody inside the chloride cells should also be considered. A first possibility is the binding of the antibody to glycoproteins as they are known to be present in chloride cells (Mansoni and Garcia-Romeu, 1972).

They are synthesized in the Golgi apparatus and are transported: 1) to the baso-lateral membrane via the tubular reticulum and 2) apically by the vesiculo-tubular system to the apical cavity (Pisam *et al.*, 1983). Their exact function is not known but it has been suggested that they could be ion-exchangers which participate in ionic regulation (Philpott, 1965). Furthermore, it has been shown from ultrastructural studies of the glycocalyx of gill cells that differences exist in the thickness (Powell *et al.*, 1994) and more importantly in the composition (Pisam *et al.*, 1980) of the cell coat of chloride cells to that of pavement cells, mucocytes and adjacent cells. However, the non-specific staining observed in red blood cells suggests that the binding of the antibody to glycoproteins is rather unlikely, as red blood cells do not possess such a structure but are more likely to possess functional AQP3, as in human erythrocytes (Roudier *et al.*, 1998). It is also worth noting that in the eel intestine, the AQP3 antibody strongly stained intra-epithelial macrophage-like bodies that are likely to possess a high solute permeability.

Another possible explanation for the binding of the AQP3 antibody in the chloride cells and in particular for its apical binding could be related to the transport of small solutes as AQP3 is not only known for transporting water but also urea and glycerol (Deen and van Os, 1998; Borgnia *et al.*, 1999; Verkman and Mitra, 2000). In teleost fish, nitrogenous waste products such as urea and ammonia can be excreted by extra-renal routes such as the gills (Mansoni and Payan, 1974; Isaia, 1982). In ureotelic fish species branchial urea clearance is believed to take place through vesicular trafficking in pavement cells (Laurent *et al.*, 2001), while urea clearance in ammoniotelic teleost fish such as the European eel, is believed to take place through the chloride cells (Mansoni and Garcia-Romeu, 1972). Furthermore, the three-fold decrease in branchial urea excretion in SW- compared to FW-acclimated eels (Mansoni and Payan, 1974) also correlates with the three-fold down-regulation of AQP3 protein expression level observed in the gills of SW- compared to FW-acclimated eels. Altogether, the presence of AQP3 in the baso-lateral network and within the apical vesiculo-tubular network of the chloride cells along with the morphological characteristics and the chemical composition of these highly specialized cells, give evidence that water can be transported through the chloride cells and, furthermore, support the hypothesis of an apical flow of solutes within these cells.

6. Conclusion

Water is the major constituent of all living cells and crosses hydrophobic plasma membranes either by simple diffusion or through a facilitative transport mechanism mediated by special protein aquaporins. Water transport across cell membranes is therefore of crucial significance especially for physiological processes that involve fluid transport. However, because of the difficulties associated with the measurements of water fluxes across teleost osmoregulatory tissues, only a few studies have investigated the nature of this essential component of the hydromineral balance which allows euryhaline teleost species to inhabit aquatic environments with widely varying salinities. Nevertheless, using the eel as a model, a major step forward has been initiated in the molecular identification and characterization of proteins associated with the water flux pathway. This should enable the determination of the role played by these aquaporins at the molecular, cellular and physiological levels as well as to allow investigations into the role of hormonal factors concerned with their regulation.

References

Alves, P., Soveral, G., Macey, R.I. and Moura, T.F. (1999) Kinetics of water transport in eel intestinal vesicles. *J. Membrane Biol.* **171**: 177–182.

Ando, M. (1975) Intestinal water transport and chloride pump in relation to seawater adaptation of the eel, *Anguilla japonica. Comp. Physiol. Biochem.* **52A**: 229–233.

Ando, M. (1980) Chloride-dependent sodium and water transport in the seawater eel intestine. *J. Comp. Physiol. B* **138:** 87–91.

Ando, M. (1981) Effects of ouabain on chloride movements across the seawater eel intestine. *J. Comp. Physiol. B* **145:** 73–79.

Ando, M. (1985) Relationship between coupled Na^+-K^+-Cl^- transport and water absorption across the seawater eel intestine. *J. Comp. Physiol. B* **155:** 311–317.

Ando, M. and Kobayashi, M. (1978) Effects of stripping of the outer layers of the eel intestine on salt and water transport. *Comp. Physiol. Biochem.* **61A:** 497–501.

Bentley, B.J. (1971) *Endocrines and Osmoregulation.* Springer, Berlin, Heidelberg, New York.

Borgnia, M., Nielsen, S., Engel, A. and Agre, P. (1999) Cellular and molecular biology of the aquaporin water channels. *Annu. Rev. Biochem.* **68:** 425–458.

Cutler, C.P. and Cramb, G. (2000) Water transport and aquaporin expression in fish. In: *Molecular Biology and Physiology of Water and Solute Transport* (eds. S. Hohmann and S. Neilsen). Kluwer Academic/Plenum Publishers, New York, pp. 433–441.

Deen, P.M.T. and van Os, C.H. (1998) Epithelial aquaporins. *Curr. Opin. Cell Biol.* **10:** 435–442.

Echevarria, M., Windhager, E.E., Tate, S.S. and Frindt, G. (1994) Cloning and expression of AQP 3, a water channel from the medullary collecting duct of rat kidney. *Proc. Natl Acad. Sci. USA.* **91:** 10997–11001.

Echevarria, M., Windhager, E.E. and Frindt, G. (1996) Selectivity of the renal collecting duct water channel Aquaporin-3. *J. Biol. Chem.* **271:** 25079–25082.

Funaki, H., Yamamoto, T., Koyama, Y., Kondo, D., Yaoita, E., Kawasaki, K., Kobayashi, H., Sawaguchi, S., Abe, H. and Kihara, I. (1998) Localization and expression of AQP5 in cornea, serous salivary glands, and pulmonary epithelial cells. *Am. J. Physiol.* **275:** C1151–C1157.

Heyman, J.B. and Engel, A. (1999). Aquaporins: phylogeny, structure, and physiology of water channels. *News Physiol. Sci.* **14:** 187–193.

Hirano, T. and Mayer-Gostan, N. (1976) Eel esophagus as an osmoregulatory organ. *Proc. Natl Acad. Sci USA.* **73:** 1348–1350.

Isaia, J. (1982) Effects of environmental salinity on branchial permeability of rainbow trout *Salmo gairdneri. J. Physiol.* **326:** 297–307.

Isaia, J. (1984) Water and nonelectrolyte permeation. In: *Fish Physiology. Volume X. Gills. Part B. Ion and Water Transfer* (eds W.S. Hoar and D.J. Randall). Academic Press, New York, London, pp. 1–38.

Ishibashi, K., Kuwahara, M., Gu, Y., Kageyama, Y., Tohsaka, A., Suzuki, F., Marumo, F. and Sasaki, S. (1997a) Cloning and functional expression of a new water channel abundantly expressed in the testis permeable to water, glycerol and urea. J. *Biol. Chem.* **272:** 20782–20786.

Ishibashi, K., Kuwahara, M., Gu, Y., Kageyama, Y., Tohsaka, A., Marumo, F. and Sasaki, S. (1997b). Cloning and functional expression of a second new aquaporin expressed in testis. *Biochem. Biophys. Res. Commun.* **237:** 714–718.

Ishibashi, K., Kuwahara, M., Gu, Y., Tanaka, Y., Marumo, F. and Sasaki, S. (1998) Cloning and functional expression of a new aquaporin (AQP 9) abundantly expressed in the peripheral leukocytes permeable to water and urea, but not glycerol. *Biochem. Biophys. Res. Commun.* **244:** 268–274.

Ishibashi, K., Kuwahara, M., Kageyama, Y., Sasaki, S., Suzuki, F. and Imai, M. (2000) Molecular cloning of a new aquaporin superfamily in mammals. In: *Molecular Biology and Physiology of Water and Solute Transport* (eds S. Hohmann and S. Neilsen). Kluwer Academic/Plenum Publishers, New York, pp. 123–125.

King, L.S. and Agre, P. (1996) Pathophysiology of the aquaporin water channels. *Annu. Rev. Physiol.* **58:** 619–648.

Kirsch, R. (1978) Role of the oesophagus in osmoregulation. In: *Osmotic and Volume Regulation. Alfred Benzon Symposium XI, Munksgaard* (eds C. B. Jorgenson and E. Skadhauge). Academic Press, New York, pp. 138–154.

Kirsch, R., Guinier, D. and Meens, R. (1975) L'équilibre hydrique de l'Anguille européenne (*Anguilla anguilla* L.). Etude du rôle de l'oesophage dans l'utilisation de l'eau de boison et étude de la perméabilité osmotique branchiale. *J Physiol (Paris)* **70:** 605–626.

Koyama, Y., Yamamoto, T., Tani, T., Nihei, K., Kondo, D., Funaki, H., Yaoita, E., Kawasaki, K., Sato, N., Hatakeyama, K. and Kihara, I. (1999) Expression and localization of aquaporins in rat gastrointestinal tract. *Am. J. Physiol.* **276:** C621–C627.

Laurent, P. and Kirsch, R. (1975) Modifications structurales de l'oesophage liées a l'osmorégulation chez l'Anguille. *C. R. Acad. Sci. (Paris)* **280:** 2227–2229.

Laurent, P., Wood, C.M., Wang, Y., Perry, S.F., Gilmour, K.M., Chevalier, C., West, M. and Walsh, P.J. (2001). Intracellular vesicular trafficking in the gill epithelium of urea excreting fish. *Cell. Tiss. Res.* **303:** 197–210.

Mansoni, A. and Garcia-Romeu, F. (1972) Accumulation et excrétion de substances organiques par les cellules à chlorure de la branchie d'Anguilla anguilla L. adaptée à l'eau de mer. *Z. Zellforsch. Mikrosk. Anat.* **133:** 389–398.

Mansoni, A. and Payan P. (1974) Urea, inulin and para-amino-hippuric acid (PAH) excretion by the gills of the eel *Anguilla anguilla L. Comp. Biochem. Physiol.* **47:** 1241–1244.

Matsuzaki, T., Suzuki, T., Koyama, H., Tanaka, S. and Takata, K. (1999) Water channel protein AQP3 is present in epithelia exposed to the environment of possible water loss. *J. Histochem. Cytochem.* **47:** 1275–1286.

Matsuzaki, T., Suzuki, T. and Takata, K. (2000). Water channel protein, aquaporin 3, in epithelial cells. In: *Molecular Biology and Physiology of Water and Solute Transport* (eds S. Hohmann and S. Neilsen). Kluwer Academic/Plenum Publishers, New York, London, pp. 167–171.

Motais, R., Isaia, J., Rankin, J. C. and Maetz, J. (1969) Adaptive changes of the water permeability of the teleostean gill epithelium in relation to external salinity. *J. Exp. Biol.* **51:** 529–546.

Nagashima, K. and Ando, M. (1993) Characterization of esophageal desalination in the seawater eel, *Anguilla japonica*. *J. Comp. Physiol. B* **164:** 47–54.

Ogasawara, T. and Hirano, T. (1984) Effects of prolactin and environmental calcium on osmotic water permeability of the gills in the eel *Anguilla japonica*. *Gen. Comp. Endocrinol.* **53:** 315–324.

Park, J.H. and Saier, Jr M.H. (1996) Phylogenetic characterization of the MIP family of transmembrane channel proteins. *J. Membrane Biol.* **153:** 171–180.

Parmelee, J.T. and Renfro, J.L. (1983) Esophageal desalination of seawater in flounder: role of active sodium transport. *Am. J. Physiol.* **245:** R888–R893.

Philpott, C.W. (1965) Halide localization in the teleost chloride cell and its identification by selected area electron diffraction. *Protoplasma* **60:** 7–23.

Pisam, M., Sardet, C. and Maetz, J. (1980) Polysaccharidic material in chloride cell of teleostean gill: modifications according to salinity. *Am. J. Physiol.* R213–R218.

Pisam, M., Chrétien, M., Rambourg, A. and Clermont, Y. (1983) Two anatomical pathways for the renewal of surface glycoproteins in chloride cells of fish gills. *Anat. Rec.* **207:** 385–397.

Potts, W.T.W. (1984) Transepithelial potentials in fish gills. In: *Fish Physiology. Volume X. Gills. Part B. Ion and Water Transfer* (eds W.S. Hoar and D.J. Randall). Academic Press, New York, London, pp. 105–128.

Powell, M.D., Speare, D.J. and Wright, G.M. (1994) Comparative ultrastructural morphology of lamellar epithelial, chloride and mucous cell, glycocalyx of the rainbow trout (*Oncorhynchus mykiss*) gill. *J. Fish Biol.* **44:** 725–730.

Preston, G.M. and Agre, P. (1991) Molecular cloning of the red cell integral protein of Mr 28,000: a member of an ancient channel family. *Proc. Natl Acad. Sci USA.* **88:** 11110–11114.

Preston, G.M., Smith, B.L., Zeidel, M.L., Moulds, J.J. and Agre, P. (1994) Mutations in aquaporin-1 in phenotypically normal humans without functional CHIP water channels. *Science* **265:** 1585–1587.

Raina, S., Preston, G.M., Guggino, W.B. and Agre, P. (1995) Molecular cloning and characterisation of an aquaporin cDNA from salivery, lacrimal, and respiratory tissues. *J. Biol. Chem.* **270:** 1908–1912.

Rankin, J.C. and Bolis, L. (1984) Hormonal control of water movement across the gills. In: *Fish Physiology. Volume X. Gills. Part B. Ion and Water Transfer* (eds W.S. Hoar and D.J. Randall). Academic Press, New York, London, pp. 177–201.

Roudier, N., Verbavatz, J-M., Maurel, C., Ripoche, P. and Tacnet, F. (1998) Evidence for the presence of aquaporin-3 in human red blood cells. *J. Biol. Chem.* **273:** 8407–8412.

Sardet, C., Pisam, M. and Maetz, J. (1979) The surface epithelium of teleostean fish gills. Cellular and junctional adaptations of chloride cells in relation to salt adaptation. *J. Cell Biol.* **80:** 96–117.

Sasaki, S., Ishibashi, K. and Marumo, F. (1998) Aquaporin-2 and -3: representatives of two subgroups of the aquaporin family colocalised in the kidney collecting duct. *Annu. Rev. Physiol.* **60:** 199–220.

Skadhauge, E. (1969) The mechanism of salt and water absorption in the intestine of the eel (*Anguilla anguilla*) adapted to waters of various salinities. *J. Physiol.* **204:** 135–158.

Skadhauge, E. (1974) Coupling of transmural flows of NaCl and water in the intestine of the eel (*Anguilla anguilla*). *J. Exp. Biol.* **60:** 535–546.

Tesch, F.-W. and Greenwood, P.H. (1977) *The Eel. Biology and Management of Anguillid Eels.* Chapman & Hall, London.

Utida, S., Hirano, T., Oide, H., Ando, M., Johnson, D.W. and Bern, H.A. (1972) Hormonal control of the intestine and urinary bladder in teleost osmoregulation. *Gen. Comp. Endo.* **S3:** 317–327.

Verkman, A.S. and Mitra, A.K. (2000) Structure and function of aquaporin water channels. *Am. J. Physiol.* **278:** F13–F28.

Yamamoto, M. and Hirano, T. (1978) Morphological changes in the esophageal epithelium of the eel, *Anguilla japonica*, during adaptation to seawater. *Cell Tiss. Res.* **192:** 25–38.

Yang, B. and Verkman, A.S. (1997) Water and glycerol permeabilities of aquaporins 1–5 and MIP determined quantitatively by expression of epitopetagged constructs in *Xenopus* oocytes. *J. Biol. Chem.* **272:** 16140–16146.

Zeuthen, T. (1995) Molecular mechanisms for passive and active transport of water. *Int. Rev. Cytol.* **160:** 99–161.

5

Hormonal control of drinking in eels: an evolutionary approach

Y. Takei

1. Introduction

Vertebrates take up water from the environment by various means. Truly terrestrial vertebrates including mammals, birds and reptiles ingest water only by oral intake or drinking. Because of their importance, the mechanisms regulating drinking have a history of research of more than a century, and the results obtained basically from mammals have been reviewed in several extensive monographs (Peters *et al.*, 1975; Fitzsimons, 1979; Rolls and Rolls, 1982; de Caro *et al.*, 1984; Grossman, 1990; Ramsay and Booth, 1991). In contrast to the truly terrestrial species, oral drinking appears to be negligible in semi-aquatic amphibians (Bentley and Yorio, 1979). Instead, they normally absorb water via the ventral skin, a process called 'cutaneous drinking' (Hillyard *et al.*, 1998).

The relative importance of oral drinking in body fluid regulation is highly diverse in aquatic fish including cyclostomes. The extant hagfish that live exclusively in the sea throughout their life have plasma ionic concentrations almost identical to environmental sea-water (SW) except for divalent cations and anions (Rankin, 1997). Thus, the movement of water and ions across the body surfaces is relatively small and thus they usually drink scarcely. However, another cyclostome species, the lamprey, that migrates between inland rivers and the sea during its life span has plasma ionic concentrations (osmolality) approximately one-third of SW. Therefore, they have well-developed mechanisms for regulation of drinking that allow them to change drinking rate in different osmotic media. This topic has been reviewed in Chapter 1 of this volume. Marine cartilaginous fish, elasmobranchs and holocephalans, accumulate urea and trimethylamine oxide (TMAO) in plasma to increase its osmolality slightly higher than SW (Hazon *et al.*, 1997). Thus, they do not lose water osmotically across the body surfaces and thus drink little water (Kobayashi *et al.*,

Osmoregulation and Drinking in Vertebrates, edited by N. Hazon & G. Flik.
© 2002 BIOS Scientific Publishers Ltd, Oxford

1983). However, they do possess the mechanisms regulating drinking, which have been reviewed in Chapter 2 of this volume.

Bony fish, teleost fish in particular, are most prosperous species that thrive in the current fresh waters (FW) and seas (Nelson, 1994). They maintain their plasma osmolality at approximately one-third that of SW in either FW or SW as observed in tetrapods and lampreys. In FW, teleost fish drink little water and urinate abundantly to cope with excess water that enters across the gill epithelia according to an osmotic gradient between plasma and media (Evans, 1993). In SW however, they literally 'drink like a fish' to compensate for water that is lost through the body surfaces by osmosis. The oral drinking is the sole means for SW fish to ingest water from the environment, as is the case for truly terrestrial animals. Euryhaline species that migrate between FW and SW or inhabit estuaries with changing environmental salinity serve as a good model for investigating the regulatory mechanisms of drinking, since they have to regulate drinking in their natural habitat when they encounter changes in environmental salinities.

The role of oral drinking in adaptation to hyper-osmotic environments has been evaluated for euryhaline teleost fish (Maetz and Skadhauge, 1968). In eels, for instance, if water ingested orally is drained through oesophageal fistula after transfer from FW to SW, plasma Na^+ concentration and haematocrit increase linearly (*Figure 1*), demonstrating gradual development of hyper-natremia and hypo-volaemia (Takei *et al.*, 1998). Consequently, the fish usually die within 5 days after transfer. However, if drained water is reintroduced into the stomach, changes in plasma Na^+ concentration and haematocrit are similar to that of intact fish (*Figure 1*). With respect to plasma Na^+ concentration, it increases abruptly soon after SW transfer with a peak within a day, and a gradual decrease follows to an equilibrium that is around 10–20% higher than the level of FW fish (*Figure 1*). These data clearly show that oral drinking is essential for teleost fish to maintain water and electrolyte balance in SW. It is obvious that Na^+ ions in the ingested SW are absorbed copiously by the intestine, because the initial increase in plasma Na^+ concentration is greater in 'reintroduced' eels than in 'drained' fish (*Figure 1*). Since the increase in plasma Na^+ concentration changes to a decrease more than 24 h after SW transfer, branchial chloride cells should be reorganized and start secreting Na^+ and Cl^- within this period.

Since the plasma of teleost fish has much lower osmolality than SW, fish are faced with a constant threat of dehydration even more seriously than terrestrial animals. For instance, if semi-aquatic toads are kept in dry air, they decrease body weight slowly, showing that the dehydration is relatively small. However, if toads are soaked in SW, they lose their body weight rapidly and die within a day because of severe hypo-volaemia and hyper-natremia. The major site of water loss is the ventral skin that serves as a respiratory epithelium of amphibians (Hillyard *et al.*, 1998). The gills serve as respiratory epithelia in teleost fish, and the surface area of the gills is much larger than that of amphibian skin (Byczkowska-Smyk, 1958). Therefore, teleost fish must be in need of retaining water in the body more desperately than amphibians when they are in hyper-osmotic environments. On the other hand, the respiratory loss of water from the lungs of truly terrestrial animals is even more reduced than from the skin of amphibians. Thus, it seems that the relative importance of drinking in maintaining water balance is greater in SW teleosts than in terrestrial animals.

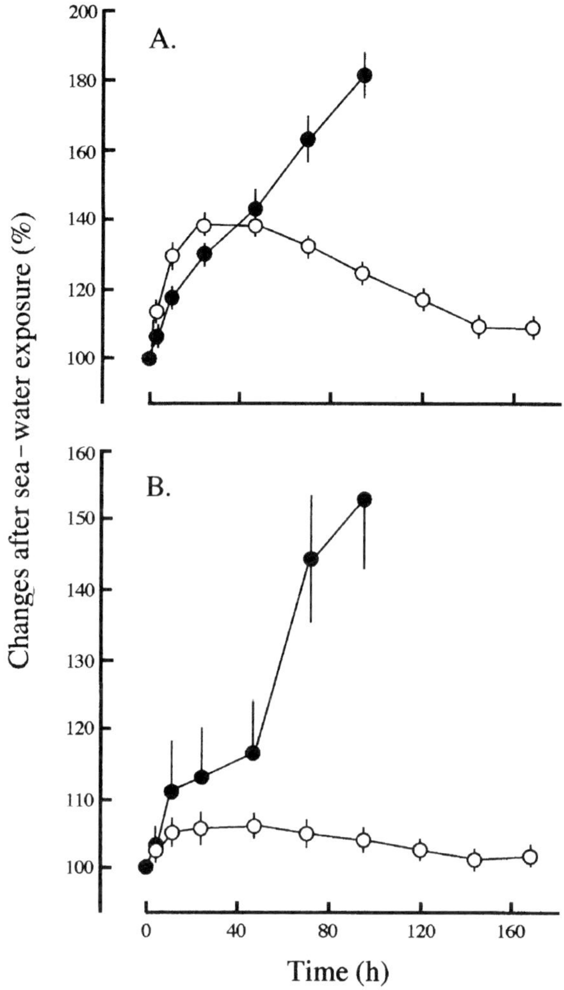

Figure 1. *Time-course of changes in (A) plasma Na^+ concentration and (B) haematocrit after transfer from FW to SW in oesophagus-cannulated eels in which ingested water was reintroduced into the stomach (○) and those in which ingested water was drained (●). The drained eels died in around 5 days (120 h) after transfer to SW.*

2. Mechanisms regulating drinking in vertebrates

2.1 *Terrestrial animals*

If terrestrial animals are deprived of water, their plasma osmolality increases while their blood volume decreases in proportion to the duration of deprivation (Fitzsimons, 1998). The decrease in blood volume is a powerful stimulus for renin secretion (Kobayashi and Takei, 1996). Thus, plasma angiotensin II (ANG II) concentration increases after water deprivation. These results indicate that an increase in plasma osmolality, a decrease in blood volume, and ANG II are involved in the regulation of drinking as demonstrated in mammals (*Figure 2*). More recently, atrial

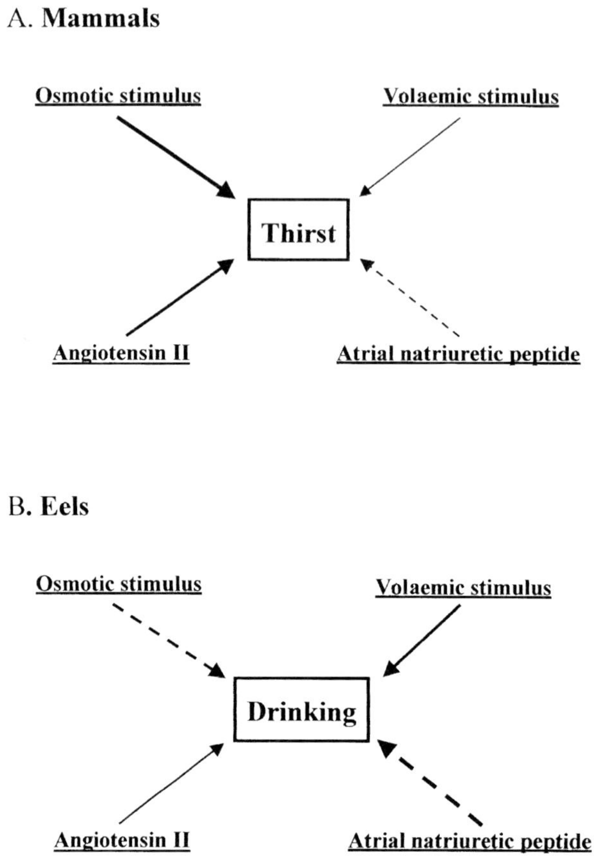

Figure 2. *Factors regulating thirst in mammals (A) and drinking in eels (B). It is not known whether eels are thirsty when they drink. Solid lines show stimulation and broken lines show inhibition. Thickness of the lines indicates strength of the regulation.*

natriuretic peptide (ANP) was shown to be an antidipsogenic hormone in mammals (*Figure 2*), as predicted from the fact that the plasma ANP concentration decreases after water deprivation.

The osmotic, volaemic and ANG II stimuli are also dipsogenic in other truly terrestrial species such as birds and reptiles. In these truly terrestrial animals, cellular dehydration of osmosensitive neurons in the hypothalamus caused by osmotic stimuli is the strongest stimulus to elicit drinking (Elkington and Taffel, 1942; Takei *et al.*, 1988a) (*Figure 2*). The potent osmotic effect is assessed by the fact that vigorous drinking occurs after the stimulus even though the other two dipsogenic stimuli, the volaemic stimulus and ANG II, are suppressed because of a simultaneous increase in blood volume (*Figure 3*). By contrast, hypovolaemia is only a weak dipsogenic stimulus in mammals and birds even though plasma concentration of ANG II, which is by itself a strong stimulus for drinking when injected directly into the brain, increases concomitantly (*Figure 2*). All stimuli induce drinking in selected species of reptiles, but the response is slow and less potent (Fitzsimons and Kaufman, 1977; Kobayashi *et al.*, 1979).

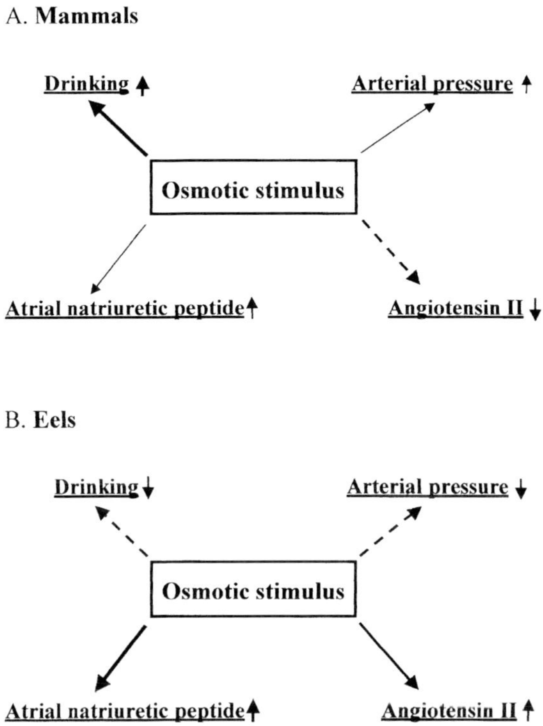

Figure 3. *Effects of osmotic stimulus on drinking, arterial pressure, plasma ANP and ANG II concentrations in mammals (A) and eels (B). Solid lines show stimulation and broken lines show inhibition. Thickness of the lines indicates strength of the regulation.*

In semi-aquatic amphibians, neither osmotic and volaemic stimuli nor ANG II causes oral drinking (Shoemaker and Nagy, 1977; Hirano *et al.*, 1978). However, these stimuli may induce thirst to initiate 'cutaneous drinking behaviour' as demonstrated for ANG II in the desert-dwelling toad (Hillyard *et al.*, 1998). ANG II also stimulates water and Na^+ absorption by the toad skin (Coviello and Brauckmann, 1973; Coviello *et al.*, 1976). It is interesting to note that the frogs start drinking when their external medium is exchanged from FW to SW (Bentley and Schmidt-Nielsen, 1971; Hirano *et al.*, 1978). This response was named 'chloride response' when it was first discovered in the eel (Hirano, 1974), which will be dealt with in some detail below. Although all the stimuli were not dipsogenic in amphibians, they most probably increase plasma vasotocin concentration as reported in mammals (Nouwen and Kuhn, 1985). Thus, these stimuli would eventually facilitate cutaneous uptake of water through an action of vasotocin (Shoemaker and Nagy, 1977; Jorgensen, 1997).

ANP is the sole antidipsogenic hormone established to date in mammals (*Figure 2*), although other humoral factors are also involved in fine-tuning of the inhibition of drinking (Fitzsimons, 1998). ANP acts on the brain to inhibit thirst and sodium appetite in the rat (Antunes-Rodrigues *et al.*, 1985; Antunes-Rodrigues *et al.*, 1986). ANP has not been identified in birds and reptiles, but mammalian ANP was shown to be dipsogenic in the quail (Okawara *et al.*, 1986). In amphibians, homologous

ANP given alone does not inhibit water and Na^+ absorption by the skin but inhibits the enhanced absorption by vasotocin in the bullfrog (Uchiyama *et al.*, 1998). It has not been examined whether ANP inhibits thirst-motivated cutaneous drinking behaviour or not.

2.2 *Aquatic animals*

In teleost fish, osmotic stimulus given as an intra-vascular injection of hypertonic NaCl solution unexpectedly inhibits drinking in the eel (Takei *et al.*, 1988c) (*Figure 2*). This is in sharp contrast to the observation in mammals and birds where the osmotic stimulus is the most potent dipsogenic stimulus. It was also shown that plasma ANG II concentration increased after osmotic stimulus (*Figure 3*), which is also opposite to the response in mammals and birds where the hyper-natremic stimulus inhibits renin secretion by way of increased Na^+ load to the macula densa (Takei *et al.*, 1988a; Kobayashi and Takei, 1996). This result suggests a lack of the macula densa in this teleost fish. Thus, the osmotic stimulus is inhibitory to drinking in the eel although dipsogenic ANG II increases simultaneously.

In contrast to the weak dipsogenic potency of osmotic stimulus, volaemic stimulus given by blood withdrawal consistently stimulates drinking in the eel (Hirano, 1974; Takei, 2000b) (*Figure 2*). The strong dipsogenic potency of the volaemic stimulus in eels is partly accounted for by a concomitant increase in plasma ANG II concentration (Takei *et al.*, 1988c; Nishimura and Madey, 1989). In mammals, however, hypo-volaemia is only a weak dipsogenic stimulus even though plasma ANG II concentration increases similarly (Takei, 2000b). The dipsogenic effect of ANG II is apparently less potent in the eel and other teleost fish than in mammals (*Figure 2*). ANP is shown to be a powerful antidipsogenic hormone in the eel (Tsuchida and Takei, 1998) (*Figure 2*). Comparing the potency between antidipsogenic ANP and dipsogenic ANG II, ANP is two orders of magnitude more potent than ANG II in the eel (Takei and Balment, 1993). Therefore, the relative potency of dipsogenic ANG II and antidipsogenic ANP is reversed between eels and mammals.

Collectively, the major differences in the regulation of drinking between terrestrial and aquatic animals are two-fold: (1) while an increase in plasma osmolality is a major dipsogenic stimulus in mammals, a decrease in blood volume is a more potent stimulus to elicit drinking in eels, and (2) while the dipsogenic potency of ANG II is greater in mammals than in eels, the antidipsogenic potency of ANP is greater in eels than in mammals (*Figure 2*). These topics will be discussed below in Section 6.

3. Renin–angiotensin system (RAS)

The renin–angiotensin cascade is initiated by the release of renin from the juxtaglomerular (JG) cells of the kidney, and the released renin acts on the plasma protein, angiotensinogen, to cleave angiotensin I (ANG I) from its N-terminus (Raizada *et al.*, 1993). Two amino-acid residues are cleaved from the C-terminus of ANG I by angiotensin converting enzyme (ACE) to form ANG II. The ANG II formation occurs mostly in the lung of terrestrial animals and the gill of aquatic animals because ACE is embedded densely in the endothelial cells of these highly vascularized organs (Kobayashi and Takei, 1996). ANG II is an active principle of the RAS. However,

N-terminally truncated ANG III and C-terminally truncated ANG IV may also have some biological functions in peripheral and/or central tissues.

3.1 *Comparative biochemistry*

The RAS has long been thought to have evolved for the first time in bony fish (Nishimura *et al.*, 1970). However, ANG I has been isolated from incubates of plasma and kidney extracts of an elasmobranch fish, *Triakis scyllia* (Takei *et al.*, 1993), and more recently from two species of lampreys, *Lampetra fluviatilis* and *Petromyzon marinus* (Y. Takei and J.C. Rankin, unpublished data). Furthermore, ANG II molecules have been biochemically identified in two species of leech, *Theromyzon tessulatum* and *Erpobdella octoculata* (Laurent *et al.*, 1995; Salzet *et al.*, 1995), and ACE has been cloned from the cDNA library of the fruit fly, *Drosophila melanogaster* (Cornell *et al.*, 1995). Since a whole set of the RAS appears to be present in protostomes, the RAS is most likely a phylogenetically ancient hormonal system that may have existed before the divergence of protostomes and deuterostomes.

3.2 *Effects on water and electrolyte balance*

ANG II was initially recognized as a vasoactive hormone as its name indicates. However, evidence is accumulating that ANG II is more deeply involved in the regulation of water and electrolyte balance (Kobayashi and Takei, 1996). It is worth noting that ANG II injected into the terrestrial slug, *Limax masimus*, induces water absorption through the foot, called 'contact rehydration' (Makara and Prior, 1985). ANG II-amide isolated from the central nervous system of the leech *Erpobdella octoculata* was shown to exert diuretic effects in this species (Salzet *et al.*, 1995). Thus, ANG II seems to be involved already in water and electrolyte homeostasis in protostome invertebrates.

In addition to the potent dipsogenic effect, ANG II induces sodium appetite to increase the oral intake of Na^+ ions (*Figure 4*). Furthermore, ANG II stimulates the absorption of ingested water and Na^+ ions by the intestine. With respect to the effects on the excretion, ANG II acts on the kidney at the level of glomeruli and tubules to decrease the excretion of water and Na^+. Thus, ANG II is a volume-retaining hormone by increasing the intake and decreasing the loss of water and Na^+ (*Figure 4*). This coincides well with the fact that hypo-volaemia is the major stimulus for renin secretion (Kobayashi and Takei, 1996). In addition to these direct actions, ANG II acts on the hypothalamus to stimulate vasopressin secretion and on the adrenal glomerulosa cells to stimulate aldosterone secretion (*Figure 4*). Vasopressin and aldosterone further promote antidiuresis and antinatriuresis.

In teleost fish, ANG II induces drinking in selected euryhaline species (Takei *et al.*, 1979a; Malvin *et al.*, 1980; Carrick and Balment, 1983) (*Figure 4*). It is not known whether or not sodium appetite is induced when SW fish drink in response to ANG II. The effect of ANG II on vasotocin secretion is not evident. However, ANG II stimulates cortisol secretion in the rainbow trout, although ACTH appears to dominate ANG II as a secretagogue (Arnold-Reed and Balment, 1994). Cortisol exhibits functions as mineralocorticoid as well as glucocorticoid in teleost fish (Balment *et al.*, 1987). ANG II is antidiuretic and antinatriuretic through glomerular actions in the rainbow trout (Brown and Balment, 1997). There has been no report on the intestinal action of ANG II in teleost fish (*Figure 4*).

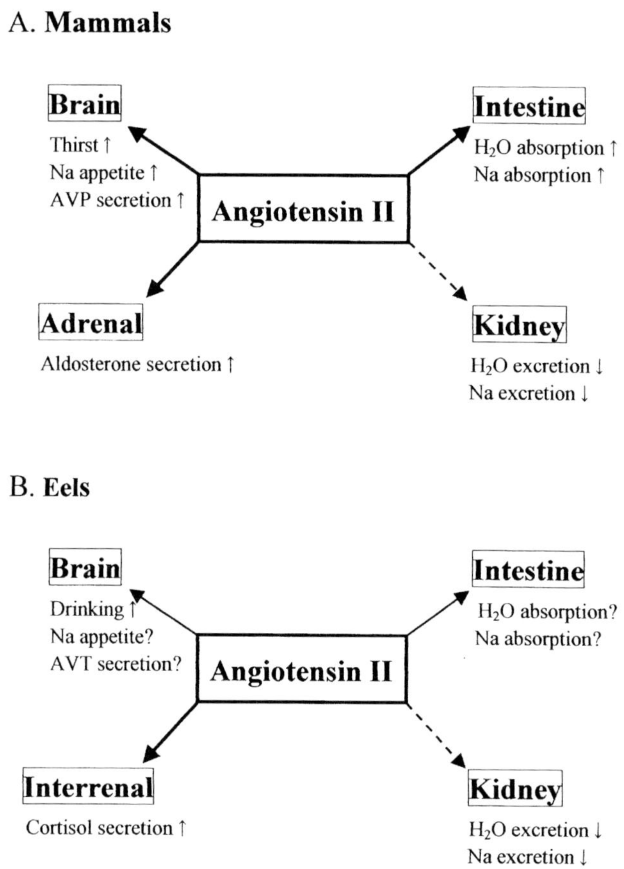

Figure 4. *Effects of ANG II on the brain, intestine, adrenal (inter-renal) gland and kidney in mammals (A) and eels (B). Solid lines show stimulation and broken lines show inhibition. Thickness of the lines indicates strength of the regulation.*

3.3 *Effects on drinking*

The dipsogenic effect of ANG II is potent in truly terrestrial animals. When injected directly into the target sites in the brain, ANG II induces drinking at a femtomole dose. ANG II induces drinking in all reptilian, avian and mammalian species examined to date, but its potency differs greatly among species that live in habitats of different water availability (Kobayashi *et al.*, 1979). Therefore, ANG II-induced thirst mechanism seems to have undergone evolution according to the ecological need for drinking. ANG II does not induce oral drinking in amphibians that do not drink in their natural habitat (Hirano *et al.*, 1978; Kobayashi *et al.*, 1979).

In teleost fish, however, ANG II is dipsogenic only at high doses that are also strongly vasoactive (pressor activity). The ANG II-induced thirst mechanism may have evolved according to the ecological need for drinking also in fish because the mechanism has been developed in euryhaline species that have to regulate drinking in their natural habitat (Kobayashi *et al.*, 1983). In stenohaline species that reside exclusively in FW or SW, ANG II generally does not induce drinking.

Site of dipsogenic action of ANG II. In the rat the site of dipsogenic action of ANG II is localized in the forebrain sites including the subfornical organ (SFO) and the antero-ventral third ventricle (AV3V) (Fitzsimons, 1998). The dipsogenic dose of ANG II at these sites is more than three orders of magnitude lower than that by peripheral injections. ANG II acts on the SFO and the medial preoptic area including the organum vasculosum of the lamina terminalis to induce drinking in the quail (Takei *et al.*, 1979b). Since these circumventricular structures are devoid of the blood–brain barrier, blood-borne ANG II can act on the neurons of these sites. However, since all components of the RAS are present in the rat brain (Ganten *et al.*, 1982), ANG II produced locally at the target sites may act in a paracrine fashion.

In the eel, it is almost impossible to inject ANG II topically to a restricted brain nucleus because of its small size. Thus, ANG II was injected into the eel cerebral ventricle. The dose required to induce drinking was 50-fold lower than that by peripheral injection (Takei, 2000b). Therefore, the site of action of ANG II may be localized in the brain also in teleost fish. However, after the forebrain and midbrain were removed, which we designated 'decerebration', a peripheral injection of ANG II still induced vigorous drinking in the eel (Takei *et al.*, 1979a). Since the SFO and AV3V are removed after 'decerebration' if they are present in fish, ANG II should have acted on other target sites located in the hindbrain. The most probable site is the medulla oblongata in which the swallowing center is localized at the bilateral areas of the reticular formation rostro-dorsally to the nucleus oliveacus inferior in mammals (Doty *et al.*, 1967). ANG II may act on the equivalent site in the medulla oblongata of the eel to induce swallowing, resulting in elicitation of drinking.

Effects of converting enzyme inhibition. To assess the role of endogenous ANG II in physiological regulation of drinking in teleost fish, ACE inhibitors have been widely used to block the conversion of ANG I to biologically active ANG II. This is because various peptide and non-peptide receptor antagonists are ineffective in teleost fish (Kobayashi and Takei, 1996), except for losartan that is effective only in the renal glomerulus of trout (Cobb and Brown, 1993).

Captopril (SQ14225) and SQ20881 are effective inhibitors for ACE in all teleost species thus far examined, so that they potently inhibit drinking induced by ANG I or by SW transfer but not by ANG II (Balment and Carrick, 1985; Beasley *et al.*, 1986; Perrott *et al.*, 1992; Tierney *et al.*, 1995; Fuentes and Eddy, 1997). Infusion of captopril into SW-adapted eels reduced plasma ANG II concentration to half, and the drinking rate and arterial pressure were decreased profoundly during infusion (Takei and Tsuchida, 2000). After infusate was changed from captopril to saline, plasma ANG II concentration rebounded above the normal level; however, the inhibition of drinking continued for more than an hour. By contrast, infusion of antiserum to ANG II reduced the plasma ANG II concentration to undetectable levels but the drinking rate and arterial pressure did not decrease significantly compared to eels infused with non-immune serum (Takei and Tsuchida, 2000). Since captopril is a fairly non-specific carboxyl dipeptidase, it should have influenced other hormonal systems and/or substrates, which may explain the potent and long-lasting inhibition of drinking. In Section 5, the involvement of bradykinin (BK) in the captopril-induced inhibition of drinking will be discussed.

4. Natriuretic peptide system (NPS)

It has long been thought that the atrium functions as a volume or stretch receptor to regulate blood volume, and the receptor function is activated by hyper-volaemia or atrial stretch (Gauer and Henry, 1963). It was in 1981 that de Bold and his colleagues reported the presence of humoral natriuretic factor in the atria, which was subsequently isolated, sequenced, and named atrial (A-type) natriuretic peptide (ANP) (de Bold, 1981). In 10 years thereafter, two other peptides, brain (B-type) natriuretic peptide and C-type natriuretic peptide (BNP and CNP) were identified as members of the natriuretic peptide (NP) family (Takei, 2000a). ANP and BNP are circulating hormones secreted from the heart, whereas CNP is principally a paracrine factor in the brain and periphery (Ruskoaho *et al.*, 1997).

4.1 *Comparative biochemistry*

In non-mammalian vertebrates, NPs have been identified by molecular biological techniques in elasmobranchs, teleost fish, amphibians and birds (Takei, 2000a). In tetrapods, ANP, BNP and CNP may be common peptides of the NP family. However, ventricular natriuretic peptide (VNP) is a teleost-specific hormone in addition to ANP and CNP. VNP is unique in that it has a long C-terminal 'tail' sequence extending from the intramolecular ring structure and that it has high affinity to both natriuretic peptide receptor-A and B (NPR-A and B) (Takei, 2000a). In elasmobranchs, CNP is the only peptide of the NP family so far identified in the brain and heart (Kawakoshi *et al.*, 2001). Very recently, a novel NP has been cloned from the heart and brain of a cyclostome, *Eptatretus burgeri*, which shares all characteristics of the NPs thus far identified (A. Kawakoshi, S. Hyodo and Y. Takei, unpublished data).

Immunoreactive ANP and ANP-like mRNA have also been identified in various invertebrate species and plants, and even in the unicellular *Paramecium multinucleatum*, using antiserum raised against mammalian ANP or using cDNA of mammalian ANP as a probe (Vesely and Giordano, 1991; Vesely and Giordano, 1992; Poulos *et al.*, 1995; Vesely and Vesely, 1999; Gehring, 1999). It is possible that the NPS is present also in invertebrates as is the case for ANG II that has been molecularly identified in annelids. However, the data for ANP are far from convincing at the present stage, and thus the molecular identification is eagerly awaited (Takei, 2001).

4.2 *Effects on water and electrolyte balance*

In mammals, the NPS is antagonistic to the RAS in every aspect of water and electrolyte regulation (Takei, 2000b) (*Figure 5*). For instance, ANP is secreted in response to an increase in blood volume, and released ANP inhibits the uptake and stimulates the excretion of water and Na^+, resulting in restoration of blood volume. In fact, ANP inhibits thirst and sodium appetite and inhibits water and NaCl absorption by the intestine (*Figure 5*). ANP also stimulates the excretion of water and Na^+ by the kidney. Furthermore, ANP inhibits the secretion of antidiuretic vasopressin and antinatriuretic aldosterone, thereby further promoting diuresis and natriuresis.

As for mammals, the whole NPS including ligands and receptors has been identified in the eel (Takei, 2000a). Physiological studies using the homologous system

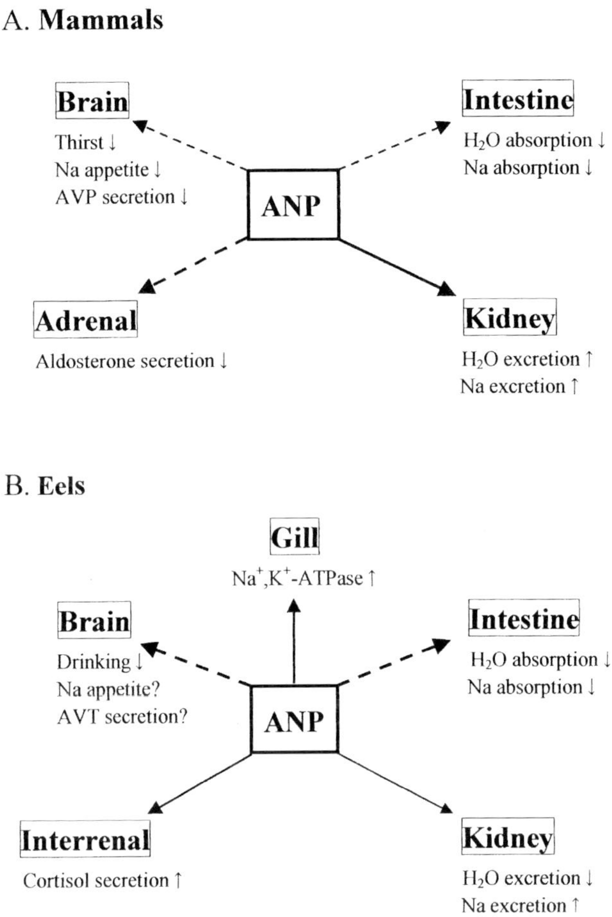

Figure 5. *Effects of ANP on the brain, intestine, adrenal (inter-renal) gland, kidney and gill (eel) in mammals (A) and eels (B). Solid lines show stimulation and broken lines show inhibition. Thickness of the lines indicates strength of the regulation.*

indicate that ANP decreases Na^+ specifically from the body and promotes SW adaptation (*Figure 5*). In fact, the ANP infusion at physiological doses decreases urine flow rate but increases urine Na^+ concentration in SW eels (Takei and Kaiya, 1998). Since arterial pressure did not decrease during the ANP infusion, the antidiuresis is not mediated by hypotension. Interestingly, the renal effect was demonstrated in SW eels but not in FW eels. Furthermore, ANP stimulates Na^+, K^+-ATPase activity in isolated gill cells from SW eels (*Figure 5*) but not in those from FW eels (Takei *et al.*, 1999). ANP also stimulates cortisol secretion in SW eels (Takei and Balment, 1993). Cortisol is known as a long-acting, SW-adapting hormone in teleost fish with synergistic actions of growth hormone (Madsen, 1990).

In contrast to the ANP actions on SW adaptation, CNP appears to be a hormone promoting FW adaptation in eels (Takei *et al.*, 2001a). CNP is circulating in large amounts in the blood of FW eels. Furthermore, CNP-specific receptors (NPR-B) are abundantly expressed in the gills and other osmoregulatory organs of FW eels

(Katafuchi *et al.*, 1994). In fact, infusion of CNP into FW eels increased plasma Na^+ concentration but CNP is not effective in SW eels (Takei *et al.*, 1999). More recently, infusion of CNP was shown to enhance the uptake of ^{22}Na from the medium in deionized water-acclimated eels (J.C. Rankin and Y. Takei, unpublished data). These results strongly suggest that CNP increases the uptake of Na^+ by the gill and promotes FW adaptation.

4.3 *Effects on drinking*

ANP potently inhibits drinking in FW and SW eels at a low dose that does not affect arterial blood pressure (Tsuchida and Takei, 1998). The antidipsogenic effect of ANP is much more potent than the dipsogenic effect of ANG II as mentioned above (Takei and Balment, 1993). In addition to the oral intake, the inhibition of intestinal water and NaCl absorption is also observed in both FW and SW eels (Ando *et al.*, 1992; Loretz and Takei, 1997). The inhibitions of drinking and intestinal NaCl absorption in SW eels certainly decrease the uptake of NaCl from the environment and thus promote SW adaptation.

As discussed above, however, drinking of SW and subsequent absorption of water by the intestine together with NaCl are essential for teleost fish to survive in a hyperosmotic SW environment (Evans, 1993). This is clearly demonstrated by the fact that if ingested water is drained through an oesophageal fistula and is not allowed to enter the stomach and intestine after transfer to SW, eels suffer from gradual hypovolaemia and hyper-natremia and die in 5 days (Takei *et al.*, 1998). Therefore, if ANP inhibits drinking and intestinal absorption of water and NaCl for a long time in SW, it would be disadvantageous for SW adaptation. However, plasma ANP concentration increases only transiently after SW transfer and its plasma level returns to a FW level in a few hours (Kaiya and Takei, 1996a) (*Figure 6*).

ANP and SW-induced drinking. It is known that a burst of drinking occurs within 1 min after transfer of eels to SW (*Figure 6*). This immediate drinking is caused by Cl^- ions in SW, because solutions that contain Cl^- salts such as choline-Cl and $MgCl_2$ are effective but $NaHCO_3$ or Na_2SO_4 are ineffective in inducing drinking when eels are exposed to these solutions (Hirano, 1974). The immediate vigorous drinking continues for about 5 min, followed by a suppression of drinking for a few hours. Thereafter, the drinking rate increases again gradually to a constant 'SW level' (Takei et al., 1988c) (*Figure 6*). This time course is just opposite to the pattern of changes in plasma ANP concentration after transfer of SW in eels (*Figure 6*). Thus, it is highly probable that ANP is involved in the physiological regulation of drinking in the initial phase of SW adaptation. The inhibition is only transient and ANP does not seem to inhibit drinking permanently in SW.

How does this transient inhibition of drinking contribute to SW adaptation? If eels continue to drink at a high rate induced by Cl^- ions after SW transfer, they would suffer from a sudden increase in plasma Na^+ concentration. In fact, the initial increase in plasma Na^+ concentration is much reduced if eels are not allowed to drink after transfer to SW (*Figure 1*), probably because Na^+ ions in SW are not absorbed by the digestive tracts (Hirano *et al.*, 1976). Taken together, it is obvious that ANP transiently inhibits drinking after eels encounter SW to diminish a sudden increase in plasma Na^+ concentration. This certainly increases the chance of survival

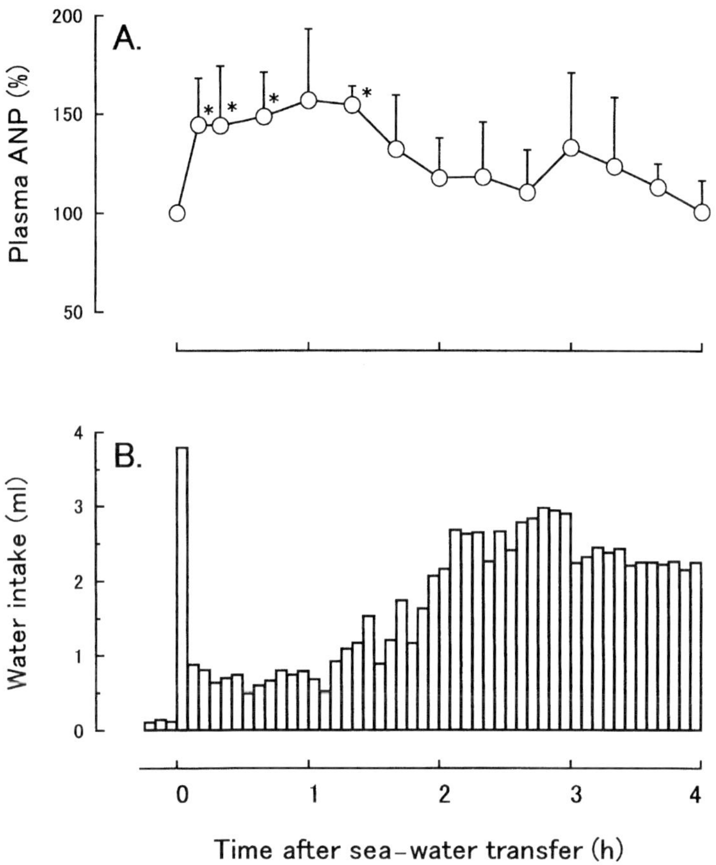

Figure 6. *Changes in plasma ANP concentrations and water intake after transfer of eels from fresh water to sea-water. *P<0.05 compared with 0-time value.*

in the initial phase of SW adaptation. It should be noted again that the ANP effects on drinking and intestinal absorption are demonstrated in FW eels as well as SW eels (Ando and Hara, 1994; Loretz and Takei, 1997). Thus, ANP can act on FW eels just after SW transfer.

ANP and antidipsogenic effect of osmotic stimulus in eels. In contrast to a potent dipsogenic action of an osmotic stimulus in mammals, it inhibited drinking in eels (Takei et al., 1988c) (*Figure 7*). More astonishingly, the same osmotic stimulus decreased arterial pressure and increased plasma ANG II concentration in eels (*Figure 3*), which is also opposite to the observation in mammals and birds (Cowley and Guyton, 1975; Takei et al., 1988a). Thus, an osmotic stimulus decreases drinking and arterial pressure in eels even though plasma ANG II increases (*Figure 3*).

The decreases in drinking rate and arterial pressure may be caused by an increase in plasma ANP concentration, since an osmotic stimulus is a powerful stimulus for ANP secretion (Kaiya and Takei, 1996b). In fact, injection of hypertonic NaCl solution increased plasma ANP concentration more than 10-fold in eels (*Figure 7*). The similar increase in plasma ANP concentration by its infusion profoundly decreased

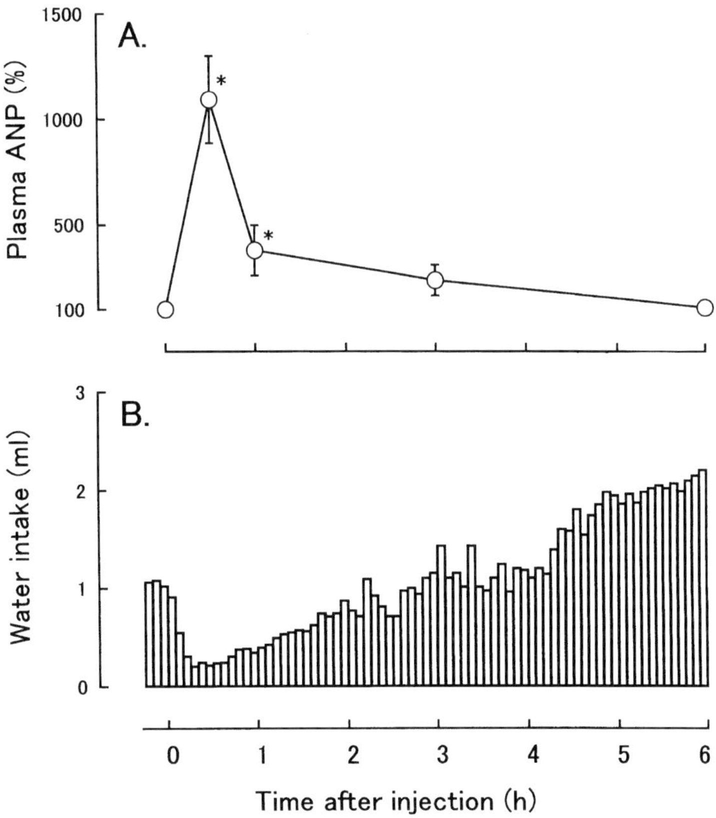

Figure 7. *Changes in plasma ANP concentrations and water intake after injection of 10% NaCl solution in eels. *P<0.05 compared with 0-time value.*

drinking rate in eels (Tsuchida and Takei, 1998). Since the patterns of an increase in plasma ANP concentration and a decrease in drinking rate are reversed, inhibition of drinking is most likely to be due to increased ANP (*Figure 7*). The inhibition by ANP should be small in mammals because an osmotic stimulus is only a weak stimulus for ANP secretion and the antidipsogenic effect of ANP is weak in mammals. Together with the possible involvement of ANP in the transient inhibition of drinking after SW transfer, it is likely that endogenous ANP is an important regulator of drinking in eels.

5. Kallikrein–kinin system (KKS)

The KKS appears to be antagonistic to the RAS because the same enzyme cleaves the active principle of the KKS, BK, to inactive metabolites and cleaves inactive ANG I to active ANG II. Actually, BK is a vasorelaxant and vasodepressor in mammals, while ANG II is vasoconstrictive and vasopressor in terms of vascular effects (Regoli *et al.*, 1997). In non-mammalian vertebrates, cardiovascular effects of BK are highly variable: vasodepressor, vasopressor or no effects were reported depending on the species examined (Conlon, 1999). In teleost fish, BK induces complicated blood pressure responses, usually tri-phasic when measured in the dorsal aorta because of

	0	1	2	3	4	5	6	7	8	9
Rat		Arg-	Pro-	Pro-	Gly-	Phe-	Ser-	Pro-	Phe-	Arg
Chicken		Arg-	Pro-	Pro-	Gly-	Phe-	Thr-	Pro-	Leu-	Arg
Alligator		Arg-	Pro-	Pro-	Gly-	Phe-	Thr-	Pro-	Phe-	Arg
Amphiuma		Phe-	Ile-	Pro-	Gly-	Leu-	Ser-	Pro-	Phe-	Arg
Lungfish		Tyr-	Gly-	Pro-	Gly-	Phe-	Ser-	Ala-	Pro-	Arg
Eel, trout, cod	Arg-	Arg-	Pro-	Pro-	Gly-	Trp-	Ser-	Pro-	Leu-	Arg
Bowfin		Arg-	Pro-	Pro-	Gly-	Trp-	Ser-	Pro-	Phe-	Arg
Sturgeon		Met-	Pro-	Pro-	Gly-	Met-	Ser-	Pro-	Phe-	Arg

Figure 8. *Amino acid sequences of BK identified to date in vertebrates.*

the involvement of various mediators (Olson *et al.*, 1997; Platzack and Conlon, 1997; Takei *et al.*, 2001b).

5.1 *Comparative biochemistry*

As observed for the RAS and NPS, the KKS appears to be a phylogenetically old hormonal system. To date, BK has been isolated in bony fish (teleosteans, holosteans, chondrosteans, and dipnoans), amphibians, reptiles, birds and mammals (Conlon, 1999). The amino acid sequences are fairly variable among tetrapods and fish, but the sequence is identical within teleostean species (trout, cod and eel) despite the fact that these three species are phylogenetically distant (*Figure 8*). In teleosts, [Arg^0]-BK that corresponds to kallidin of mammals is more bioactive than BK (Platzack and Conlon, 1997; Takei *et al.*, 2001b), but the circulating form has not been identified due to its extremely short half-life in blood.

5.2 *Effects on water and electrolyte balance*

Regarding the renal effects, BK is natriuretic and diuretic in mammals, which is again opposite to the antinatriuretic and antidiuretic effects of ANG II (Siragy, 1993; Seymour *et al.*, 1994). However, BK is dipsogenic in the rat as is ANG II when injected systemically (Fregly and Rowland, 1991). The dipsogenic effect of BK is so weak that captopril (inhibitor of kininase II) has to be administered together to ensure the effect. The effect of homologous BK on water and electrolyte balance has not been examined in non-mammalian tetrapods.

In teleost fish, intra-arterial bolus injections of homologous BK inhibit drinking dose-dependently in SW eels that drink copiously and at a constant rate (Takei *et al.*, 2001b). The minimum dose of BK to obtain an antidipsogenic effect is comparable to that of the dipsogenic effect of ANG II in the eel (Takei *et al.*, 1979a). Slow infusion of BK into the circulation also decreases drinking rate in a dose-dependent manner without changes in arterial pressure. Thus, the antidipsogenic effect of BK is highly potent in the eel, indicating that the potent antidipsogenic effect of captopril mentioned above is due in part to the increased plasma BK (Takei and Tsuchida, 2000). At the highest dose of BK infusion, drinking is profoundly inhibited despite the observation that plasma ANG II is increased. On the other hand, ANP infusion at antidipsogenic doses decreased plasma ANG II concentration (Tsuchida and Takei, 1998), whereas ANG II infusion at dipsogenic doses increased plasma ANP concentration in the eel (Tsuchida and Takei, 1999). It seems, therefore, that the regulation of drinking cannot be accounted for by a single hormone but results from complex interactions of the RAS, NPS, KKS and other hormonal systems in the eel.

6. Evolutionary consideration

In mammals, mechanisms regulating drinking have a history of research of over 200 years (Fitzsimons, 1998). It was only 30 years ago that ANG II was discovered as one of the essential dipsogenic stimuli (Epstein *et al.*, 1970). Since then, various humoral factors have been extensively pursued, but most are found to be minor factors as illustrated by BK and ANP. Compared with the ample data in mammals, the data in fish are now accumulating. However, it would be worthwhile to consider now how the mechanisms regulating drinking have evolved from aquatic fish to terrestrial animals.

6.1 *Osmotic stimulus versus volaemic stimulus*

As evidenced above, an osmotic stimulus is a major dipsogenic regulator in mammals and birds, but a volaemic stimulus is a more dominant regulator in fish. It is intriguing to consider how this difference arose. In mammals and birds, plasma Na^+ concentration appears to be the primary target of body fluid regulation. Thus, if they are deprived of water, plasma Na^+ concentration is maintained constant while blood volume decreases quickly after water deprivation (Fitzsimons, 1979; Takei *et al.*, 1988b). This may be due to the fact that increased plasma Na^+ concentration hyperpolarizes the cells and decreases the excitability of neurons. In teleost fish, a situation similar to water deprivation in terrestrial animals can be induced by transferring the euryhaline fish from FW to SW. In contrast to mammals and birds, however, blood volume is primarily maintained while plasma Na^+ concentration increases quickly as demonstrated in the eel (*Figure 1*). Therefore, it seems that blood volume is the primary target of body fluid regulation in the eel.

It is possible that an increase in ambient osmotic pressure scarcely affects the function of fish cells. It is shown that fish cells in culture are highly resistant to changes in environmental osmolality, so that they grow much better in hypertonic media than mammalian culture cells (Hashimoto *et al.*, 1998). Furthermore, teleost fish can tolerate profound increases in plasma osmolality compared with mammals and birds; eels can survive an up to 80% increase in plasma Na^+ concentration (Takei *et al.*, 1998), but the brain function is severely damaged with much smaller increases in mammals (Maxwell *et al.*, 1987). There is a possibility that fish cells adjust its volume quickly after perturbation of ambient osmolality by changing the activity of osmolyte channels/transporters (Goldstein and Davis, 1994). Thus, cellular dehydration might be suppressed in fish, which may partly explain the weak dipsogenic potency of the osmotic stimulus in eels.

It is not known at present why eels are so sensitive to the dipsogenic effect of hypo-volaemia. Because of their aquatic life, blood volume of teleost fish is vulnerable to changes in environmental salinity (Rankin and Bolis, 1984). Furthermore, blood volume of teleost fish is small, only equivalent to 3–4% of body weight, and it is more strictly maintained than plasma osmolality (Takei, 1988). Therefore, it seems that the sensitivity to the volaemic stimulus is enhanced in teleost fish to ensure volume homeostasis by drinking in response to small changes in blood volume.

6. 2 *Evolution of receptive sites for ANG II*

In mammals and birds, ANG II acts on the dipsogenic receptors in the forebrain such as SFO and AV3V to induce thirst, but ANG II appears to act on the swallowing centre in the hindbrain to induce swallowing in eels. The evolution of the receptive sites may have occurred during the evolutionary transition from aquatic to terrestrial life. Since fish live in water, they can drink only by swallowing reflex without feeling thirsty. This is the major difference from terrestrial species, which have to feel thirsty to initiate a series of drinking behaviours. Initially, thirst motivates the animals towards water to take it in their mouth. Then, subsequent swallowing completes the drinking behaviours of terrestrial animals. In fish, however, thirst-motivated behaviour can be omitted because water is always available in their mouth. It seems, therefore, that ANG II receptors initially appeared at the swallowing centre in the medulla oblongata of teleost fish; however, new receptors for ANG II have appeared as dipsogenic (thirst-inducing) receptors in the forebrain of terrestrial animals during the course of evolution towards the land. The ANG II receptors for swallowing may still be present in the medulla oblongata of the cat because injection of ANG II into the site induced swallowing (Cooling and Day, 1975).

6.3 *Relative importance of hormonal systems*

It is of interest to consider why the relative importance of the three hormonal systems, RAS, KKS and NPS, in the regulation of drinking has changed between teleost

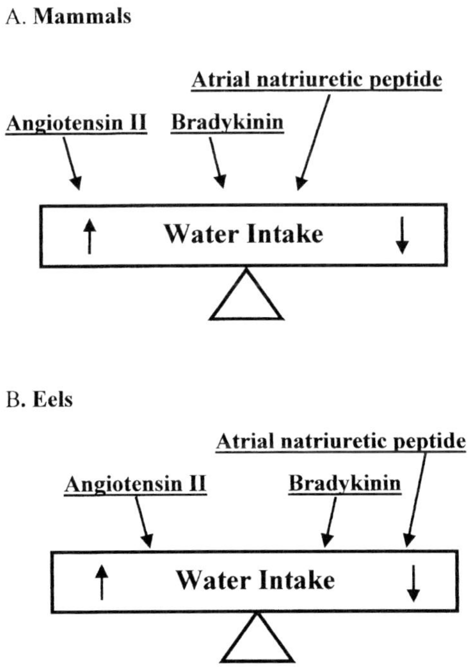

Figure 9. *Relative dipsogenic and antidipsogenic potencies of ANG II, ANP and BK in mammals (A) and eels (B). Potency is greater as the position of the arrow is farther from a fulcrum of the seesaw.*

fish (eel) and mammals (rat). It has been shown that the dipsogenic potency of ANG II is much greater in mammals than in eels, but the antidipsogenic potency of ANP is much greater in eels than in mammals (*Figure 9*). Furthermore, BK is a weak dipsogenic hormone in mammals but is a potent antidipsogenic hormone in eels. Considering these results, it is apparent that the mechanisms eliciting thirst are much more developed and dominant in mammals, whereas the mechanisms inhibiting drinking are more overwhelming in eels (*Figure 9*).

As discussed above, this difference appears also to reflect the difference in their habitat. Since mammals live on the land, they have to seek water to drink, which is ensured only by the mechanisms inducing thirst. By contrast, since eels are in water, if the mechanisms that inhibit drinking are absent, they most probably suffer from excess drinking and subsequent over-hydration. It seems, therefore, that the regulation of drinking in aquatic animals in general would be characterized by the preponderance of inhibitory mechanisms for drinking. This rule seems to apply also to the hormonal control of drinking.

References

Ando, M. and Hara, I. (1994) Alteration of sensitivity to various regulators in the intestine of the eel following seawater acclimation. *Comp. Biochem. Physiol.* **109A:** 447–453.

Ando, M., Kondo, K. and Takei, Y. (1992) Effects of eel atrial natriuretic peptide on NaCl and water transport across the intestine of the seawater eel. *J. Comp. Physiol.* **B162:** 436–439.

Antunes-Rodrigues, J., McCann, S.M., Rogers, L.C. and Samson, W.K. (1985) Atrial natriuretic factor inhibits water intake in conscious rats. *Proc. Natl Acad. Sci. USA* **82:** 8720–8724.

Antunes-Rodrigues, J., McCann, S.M. and Samson, W.K. (1986) Central administration of atrial natriuretic factor inhibits saline preference in the rat. *Endocrinology* **118:** 1726–1728.

Arnold-Reed, D.E. and Balment, R.J. (1994) Peptide hormones influence in vitro interrenal secretion in the trout, *Oncorhynchus mykiss. Gen. Comp. Endocrinol.* **96:** 85–91.

Balment, R.J. and Carrick, S. (1985) Endogenous renin–angiotensin system and drinking behaviour in flounder. *Am. J. Physiol.* **248:** R157–R160.

Balment, R.J., Hazon, N. and Perrott, M.N. (1987) Control of corticosteroid secretion and its relation to osmoregulation in lower vertebrates. In: *Comparative Physiology of Environmental Adaptations I. Adaptation to Salinity and Dehydration* (eds R. Kirsch and B. Lahlou). Karger, Basel, pp. 92–102.

Beasley, D., Shier, D.N., Malvin, R.L. and Smith, G. (1986) Angiotensin-stimulated drinking in marine fish. *Am. J. Physiol.* **250:** R1034–R1038.

Bentley, P.J. and Schmidt-Nielsen, K. (1971) Acute effects of sea water on frogs (*Rana pipiens*). *Comp. Biochem. Physiol.* **40A:** 547–548.

Bentley, P.J. and Yorio, T. (1979) Do frogs drink? *J. Exp. Biol.* **79:** 41–46.

Brown, J.A. and Balment, R.J. (1997) Teleost renal function: regulation by arginine vasotocin and by angiotensins. In: *Ionic Regulation in Animals* (eds N. Hazon, F.B. Eddy and G. Flik). Springer, Berlin, pp. 150–164.

Byczkowska-Smyk, W. (1958) The respiratory surface of the gills in teleosts. Part II. The respiratory surface of the gills in the eel (*Anguilla anguilla* L.), the loach (*Misgurnus fossilis* L.), and the perch-pike (*Lucioperca lucioperca* L.). *Acta Biol. Cravoviensia Ser. Zool.* **1:** 84–97.

Carrick, S. and Balment, R.J. (1983) The renin–angiotensin system and drinking in euryhaline flounder, *Platichthys flesus. Gen. Comp. Endocrinol.* **51:** 423–433.

Cobb, C.S. and Brown, J.A. (1993) Characterization of putative glomerular receptors for angiotensin II in the rainbow trout *Oncorhynchus mykiss* using the antagonists losartan, PD123177, and saralasin. *Gen. Comp. Endocrinol.* **92:** 123–131.

Conlon, J.M. (1999) Bradykinin and its receptors in non-mammalian vertebrates. *Regul. Pept.* **79:** 1–81.

Cooling, M.J. and Day, M.D. (1975) Drinking behaviour in the cat induced by renin, angiotensin I, II and isoprenaline. *J. Physiol.* (London) **244:** 325–336.

Cornell, M.J., Williams, T.A., Lamango, N.S., Coates, D., Corvol, P., Soubrier, F., Hoheisel, J., Lehrach, H. and Isaac, R.E. (1995) Cloning and expression of an evolutionary conserved single-domain angiotensin converting enzyme from *Drosophila melanogaster. J. Biol. Chem.* **270:** 13613–13616.

Coviello, A. and Brauckmann, E.S. (1973) Hydrosmotic effect of angiotensin II: isolated toad skin. *Acta Physiol. Latinoam.* **23:** 18–23.

Coviello, A., Elso, G. and Fernandez, F.M. (1976) Effect of angiotensin II on short-circuit current in amphibian membranes. *Biochem. Pharmacol.* **25:** 106–107.

Cowley, A.W., Jr and Guyton, A.C. (1975) Baroreceptor reflex effects on transient and steady-state hemodynamics of salt-loading hypertension in dogs. *Circ. Res.* **36:** 536–546.

de Bold, A.J., Borenstein, H.B., Veress, A.T. and Sonnenberg, H. (1981) A rapid and potent natriuretic response to intravenous injection of atrial myocardial extract in rats. *Life Sci.* **28:** 89–94.

de Caro, G., Epstein, A.N. and Massi, M. (1984) *The Physiology of Thirst and Sodium Appetite. NATO ASI Series A: Life Sciences Vol. 105.* Plenum Press, London.

Doty, R.W., Richmond, W.H. and Storey, A.T. (1967) Effects of medullary lesions on coordination of deglutition. *Exp. Neurol.* **17:** 91–106.

Elkinton, J.R. and Taffel, M. (1942) Prolonged water deprivation in the dog. *J. Clin. Invest.* **21:** 787–794.

Epstein, A.N., Fitzsimons, J.T. and Rolls, B.J. (1970) Drinking induced by injection of angiotensin into the brain of the rat. *J. Physiol. London* **210:** 457–474.

Evans, D.H. (1993) Osmotic and ionic regulation. In: *The Physiology of Fish* (ed. D.H. Evans). CRC Press, Boca Raton, FL, pp. 315–341.

Fitzsimons, J.T. (1979) *The Physiology of Thirst and Sodium Appetite.* Cambridge University Press, Cambridge.

Fitzsimons, J.T. (1998) Angiotensin, thirst, and sodium appetite. *Physiol. Rev.* **78:** 583–686.

Fitzsimons, J.T. and Kaufman, S. (1977) Cellular and extracellular dehydration, and angiotensin as stimuli to drinking in the common iguana, *Iguana iguana. J. Physiol.* (London) **265:** 443–463.

Fregly, M.J. and Rowland, N.E. (1991) Bradykinin-induced dipsogenesis in captopril-treated rats. *Brain Res. Bull.* **26:** 169–172.

Fuentes, J. and Eddy, F.B. (1997) Effect of manipulation of the renin–angiotensin system in control of drinking in juvenile Atlantic salmon (*Salmo salar* L.) in fresh water and after transfer to sea water. *J. Comp. Physiol.* **B167:** 438–443.

Ganten, D., Prinz, M., Phillips, M.I. and Scholkens, B.A. (1982) *The Renin Angiotensin System in the Brain.* Springer, Berlin.

Gauer, O.H. and Henry, J.P. (1963) Circulatory basis of fluid control. *Physiol. Rev.* **43:** 423–481.

Gehring, C.A. (1999) Natriuretic peptides—a new class of plant hormone? *Annals Botany* **83:** 329–334.

Goldstein, L. and Davis, E.M. (1994) Taurine, betaine, and inositol share a volume-sensitive transporter in skate erythrocyte cell membrane. *Am. J. Physiol.* **267:** R426–R431.

Grossman, S. P. (1990) *Thirst and Sodium Appetite. Physiological Basis.* Academic Press, San Diego.

Hashimoto, H., Takeuchi, K., Matsuo, Y., Yokoyama, Y., Toyohara, H. and Sakaguchi, M. (1998) Cell growth of fish cultures in hyper and hypoosmotic media. *Fisheries Sci.* **64:** 350–351.

Hazon, N., Tierney, M.L., Anderson, W.G., Mackenzie, S., Cutler, C. and Cramb, G. (1997) Ion and water balance in elasmobranch fish. In: *Ionic Regulation in Animals* (eds N. Hazon, F.B. Eddy and G. Flik). Springer, Berlin, pp. 70–86.

Hillyard, S.D., von Hoff, K.S. and Propper, C. (1998) The water absorption response: a behavioural assay for physiological processes in terrestrial amphibians. *Physiol. Zool.* **71:** 127–138.

Hirano, T. (1974) Some factors regulating water intake in the eel, *Anguilla japonica*. *J. Exp. Biol.* **61:** 737–747.

Hirano, T., Morisawa, M., Ando, M. and Utida, S. (1976) Adaptive changes in ion and water transport mechanism in the eel intestine. In: *Intestinal Ion Transport* (ed. J.W.L. Robinson). Medical and Technical, Lancaster, pp. 301–317.

Hirano, T., Takei, Y. and Kobayashi, H. (1978) Angiotensin and drinking in the eel and the frog. In: *Osmotic and Volume Regulation* (eds C.B. Jorgensen and E. Skadhauge). Munksgaard, Copenhagen, pp. 123–128.

Jorgensen, C.B. (1997) 200 years of amphibian water economy: From Robert Townson to the present. *Biol. Rev.* **72:** 153–237.

Kaiya, H. and Takei, Y. (1996a) Changes in plasma atrial and ventricular natriuretic peptide concentrations after transfer of eels from fresh water and seawater or *vice versa*. *Gen. Comp. Endocrinol.* **104:** 337–345.

Kaiya, H. and Takei, Y. (1996b) Osmotic and volaemic regulation of atrial and ventricular natriuretic peptide secretion in conscious eels. *J. Endocrinol.* **149:** 441–447.

Katafuchi, T., Takashima, A., Kashiwagi, M., Hagiwara, H., Takei, Y. and Hirose, S. (1994) Cloning and expression of eel natriuretic peptide receptor B (NPR-B) and its comparison with the mammalian counterparts. *Eur. J. Biochem.* **222:** 835–842.

Kawakoshi, A., Hyodo, S. and Takei, Y. (2001) CNP is the only natriuretic peptide in an elasmobranch fish, *Triakis scyllia*. *Zool. Sci.* **18:** (in press).

Kobayashi, H. and Takei, Y. (1996) *The Renin–Angiotensin System – Comparative Aspects, Zoophysiology Vol. 35*. Springer, Berlin.

Kobayashi, H., Uemura, H., Wada, M. and Takei, Y. (1979) Ecological adaptation of angiotensin-induced drinking in tetrapods. *Gen. Comp. Endocrinol.* **38:** 93–104.

Kobayashi, H., Uemura, H., Takei, Y., Itatsu, N., Ozawa, M. and Ichinohe, K. (1983) Drinking induced by angiotensin II in fishes. *Gen. Comp. Endocrinol.* **49:** 295–306.

Laurent, V., Bulet, P. and Salzet, M. (1995) A comparison of the leech *Theromyzon tessulatum* angiotensin I-like molecule with forms of vertebrate angiotensinogens: a hormonal system conserved in the course of evolution. *Neurosci. Lett.* **190:** 175–178.

Loretz, C.A. and Takei, Y. (1997) Natriuretic peptide inhibition of intestinal salt absorption in the Japanese eel: physiological significance. *Fish Physiol. Biochem.* **17:** 319–324.

Madsen, S.S. (1990) The role of cortisol and growth hormone in seawater adaptation and development of hypoosmoregulatory mechanisms in sea trout parr (*Salmo trutta trutta*). *Gen. Comp. Endocrinol.* **79:** 1–11.

Maetz, J. and Skadhauge, E. (1968) Drinking rates and gill ionic turnover in relation to external salinities in the eel. *Nature* **217:** 371–373.

Makara, M.E. and Prior, D.J. (1985) Angiotensin II can initiate contact-rehydration in terrestrial slugs. *J. Exp. Biol.* **119:** 385–388.

Malvin, R.L., Schiff, D. and Eiger, S. (1980) Angiotensin and drinking rates in the euryhaline killifish. *Am. J. Physiol.* **239:** R31–R34.

Maxwell, M.H., Kleeman, C.R. and Nairns, R.G. (1987) *Clinical Disorders of Fluid and Electrolyte Metabolism*, 4th Edn. McGraw-Hill, New York.

Nelson, J.S. (1994) *Fishes of the World*, 3rd Edn. Wiley, New York.

Nishimura, H. and Madey, A.M. (1989) Signals controlling renin release in aglomerular toadfish. *Fish Physiol. Biochem.* **7:** 323–329.

Nishimura, H., Oguri, M., Ogawa, M., Sokabe, H. and Imai, M. (1970) Absence of renin in kidneys of elasmobranchs and cyclostomes. *Am. J. Physiol.* **218:** 911–915.

Nouwen, E.J. and Kuhn, E.R. (1985) Volumetric control of arginine vasotocin and mesotocin release in the frog (*Rana ridibunda*). *J. Endocrinol.* **105:** 371–377.

Okawara, Y., Seki, K. and Kobayashi, H. (1986) Biologically active peptides and water intake II. Increases of water intake and excretion following injection of atrial natriuretic peptide in the Japanese quail, *Coturnix coturnix japonica. Zool. Sci.* **3:** 831–836.

Olson, K.R., Conklin, D.J., Weaver, L., Jr, Duff, D.W., Herman, C.A., Wang, X. and Conlon, J.M. (1997) Cardiovascular effects of homologous bradykinin in rainbow trout. *Am. J. Physiol.* **272:** R1112–R1120.

Perrott, M.N., Grierson, C.E., Hazon, N. and Balment, R.J. (1992) Drinking behaviour in sea water and fresh water teleosts, the role of the renin–angiotensin system. *Fish Physiol. Biochem.* **10:** 161–168.

Peters, G., Fitzsimons, J.T. and Peters-Haefeli, L. (1975) *Control Mechanisms of Drinking.* Springer, Berlin.

Platzack, B. and Conlon, J.M. (1997) Purification, structural characterization, and cardiovascular activity of cod bradykinin. *Am. J. Physiol.* **272:** R710–R717.

Poulos, J.E., Gower, W.R., Jr, Friedl, F.E. and Vesely, D.L. (1995) Atrial natriuretic peptide gene expression within invertebrate heart. *Gen. Comp. Endocrinol.* **100:** 61–68.

Raizada, M.K., Phillips, M.I. and Sumners, K. (1993) *Cellular and Molecular Biology of the Renin–Angiotensin System.* CRC Press, Boca Raton, FL.

Ramsay, D.J. and Booth, D.A. (1991) *Thirst. Physiological and Psychological Aspects.* Springer, London.

Rankin, J.C. (1997) Osmotic and ionic regulation in cyclostomes. In: *Ionic Regulation in Animals* (eds N.. Hazon, F.B. Eddy and G. Flik). Springer, Berlin, pp. 50–69.

Rankin, J.C. and Bolis, L. (1984) Hormonal control of water movement across the gills. In: *Fish Physiology, Vol. X. Gills, Part B. Ion and Water Transfer* (eds W.S. Hoar and D.J. Randall). Academic Press, London, pp. 177–201.

Regoli, D., Rizzi, A. and Cab, G. (1997) Pharmacology of the kallikrein–kinin system. *Pharmacol. Res.* **35:** 513–515.

Rolls, B.J. and Rolls, E.T. (1982) *Thirst.* Cambridge University Press, Cambridge.

Ruskoaho, H., Leskinen, H., Magga, J., Taskinen, P., Mantymma, P., Vuolteenaho, O. and Leppaluoto, J. (1997) Mechanisms of mechanical load-induced atrial natriuretic peptide secretion: role of endothelin, nitric oxide, and angiotensin II. *J. Mol. Med.* **75:** 876–885.

Salzet, M., Bulet, P., Wattez, C., Verger-Bocquet, M. and Malecha, J. (1995) Structural characterization of a diuretic peptide from the central nervous system of the leech *Erpobdella octoculata. J. Biol. Chem.* **270:** 1575–1582.

Seymour, A.A., Sheldon, J.H., Smith, P.L., Asaad, M. and Rogers, W.L. (1994) Potentiation of the renal responses to bradykinin by inhibition of neutral endopeptidase 3.4.24.11 and angiotensin converting enzyme in anesthetized dogs. *J. Pharmacol. Exp. Therap.* **269:** 263–270.

Shoemaker, V.H. and Nagy, K.A. (1977) Osmoregulation in amphibians and reptiles. *Annu. Rev. Physiol.* **39:** 449–471.

Siragy, H.M. (1993) Evidence that intrarenal bradykinin plays a role in regulation of renal function. *Am. J. Physiol.* **265:** E648–E654.

Takei, Y. (1988) Changes in blood volume after alteration of hydromineral balance in conscious eels, *Anguilla japonica. Comp. Biochem. Physiol.* **91A:** 293–297.

Takei, Y. (2000a) Structural and functional evolution of the natriuretic peptide system in vertebrates. *Int. Rev. Cytol.* **194:** 1–60.

Takei, Y. (2000b) Comparative physiology of body fluid regulation in vertebrates with special reference to thirst regulation. *Jpn. J. Physiol.* **50:** 171–186.

Takei, Y. (2001) Does the natriuretic peptide system exist throughout the animal and plant kingdom? *Comp. Biochem. Physiol.* **129B:** 559–573.

Takei, Y. and Balment, R.J. (1993) Biochemistry and physiology of a family of natriuretic peptide in eels. *Fish Physiol. Biochem.* **11:** 183–188.

Takei, Y. and Kaiya, H. (1998) Antidiuretic effect of eel ANP infused at physiological doses in seawater-adapted eels, *Anguilla japonica. Zool. Sci.* **15:** 399–404.

Takei, Y. and Tsuchida, T. (2000) Role of the renin–angiotensin system in drinking of seawater-adapted eels, *Anguilla japonica*: a reevaluation. *Am. J. Physiol.* **279:** R1105–R1111.

Takei, Y., Hirano, T. and Kobayashi, H. (1979a) Angiotensin and water intake in the Japanese eel, *Anguilla japonica. Gen. Comp. Endocrinol.* **38:** 446–475.

Takei, Y., Kobayashi, H., Yanagisawa, M. and Bando, T. (1979b) Involvement of catecholaminergic nerve fibers in angiotensin II-induced drinking in the Japanese quail, *Coturnix coturnix japonica. Brain Res.* **174:** 229–244.

Takei, Y., Okawara, Y. and Kobayashi, H. (1988a) Drinking induced by cellular dehydration in the quail, *Corutnix coturnix japonica. Comp. Biochem. Physiol.* **90A:** 291–296.

Takei, Y., Okawara, Y. and Kobayashi, H. (1988b) Water intake caused by water deprivation in the quail, *Corutnix coturnix japonica. J. Comp. Physiol.* **B158:** 519–525.

Takei, Y., Okubo, J. and Yamaguchi, K. (1988c) Effect of cellular dehydration on drinking and plasma angiotensin II level in the eel, *Anguilla japonica. Zool. Sci.* **5:** 43–51.

Takei, Y., Hasegawa, Y., Watanabe, T.X., Nakajima, K. and Hazon, N. (1993) A novel angiotensin I isolated from an elasmobranch fish. *J. Endocrinol.* **139:** 281–285.

Takei, Y., Tsuchida, T. and Tanakadate, A. (1998) Evaluation of water intake in seawater adaptation in eels using a synchronized drop counter and pulse injector system. *Zool. Sci.* **15:** 677–682.

Takei, Y., Flik, G. and Hirose, S. (1999) Interaction of natriuretic peptide system and the renin–angiotensin system in eel osmoregulation. In: *Recent Developments in Comparative Endocrinology and Neurobiology* (eds E.W. Roubos, S.E. Wendelaar Bonga, H. Vaudry and A. De Loof). Shakers Publishing B. V., Maastricht, pp. 144–148.

Takei, Y., Inoue, K., Ando, K., Ihara, T., Katafuchi, T., Kashiwagi, M. and Hirose, S. (2001a) Enhanced expression and release of C-type natriuretic peptide in peripheral tissues of eels, *Anguilla japonica*, adapted to fresh water. *Am. J. Physiol.* **280:** R1727–R1735.

Takei, Y., Tsuchida, T., Li, Z. and Conlon, M. (2001b) Antidipsogenic effects of eel bradykinins in the eel, *Anguilla japonica. Am. J. Physiol.* **281:** R1090–R1096.

Tierney, M.L., Luke, G., Cramb, G. and Hazon, N. (1995) The role of the renin–angiotensin system in the control of blood pressure and drinking in the European eel, *Anguilla anguilla. Gen. Comp. Endocrinol.* **100:** 39–48.

Tsuchida, T. and Takei, Y. (1998) Effects of homologous atrial natriuretic peptide on drinking and plasma ANG II level in eels. *Am. J. Physiol.* **275:** R1605–R1610.

Tsuchida, T. and Takei, Y. (1999) A potent dipsogenic action of homologous angiotensin II infused at physiological doses in eels, *Anguilla japonica. Zool. Sci.* **16:** 479–483.

Uchiyama, M., Takeuchi, T. and Matsuda, K. (1998) Effects of homologous natriuretic peptide in isolated skin of the bullfrog, *Rana catesbeiana. Comp. Biochem. Physiol.* **B120:** 37–42.

Vesely, D.L. and Giordano, A.T. (1991) Atrial natriuretic hormonal system in plants. *Biochem. Biophys. Res. Commun.* **179:** 695–700.

Vesely, D.L. and Giordano, A.T. (1992) Atrial natriuretic factor-like peptide and its prohormone within single cell organisms. *Peptides* **13:** 177–182.

Vesely, M.D. and Vesely, D.L. (1999) Environmental upregulation of the atrial natriuretic peptide gene in the living fossil, *Limulus polyphemus. Biochem. Biophys. Res. Commun.* **254:** 751–756.

6

The role of arginine vasotocin in teleost fish osmoregulation

J. Warne

1. Introduction

1.1 *Teleosts and osmoregulation*

The majority of teleost fish regulate and maintain the osmotic concentration of their body fluids to a specific set point usually between 250 and 350 mOsm kg H_2O^{-1} irrespective or whether they are in a freshwater (FW) or sea-water (SW) environment (Holmes and Donaldson, 1969). Living in either FW or SW presents a fish with very different osmoregulatory problems. In FW the fish is hyper-osmotic to its environment which means there is a tendency for the animal to lose salts and gain water. In SW the opposite is the case, the fish is hypo-osmotic to its environment and there is a tendency for fish to lose water and gain salts. The transporting epithelia in the gills, kidney/bladder and gastro-intestinal tract are responsible for maintaining the osmotic composition of body fluids and rectifying any imbalance, but the role these tissues play in each environment is very different. In FW the gills can actively take up ions to offset any losses not balanced by diet and the kidney excretes large amounts of dilute urine to eliminate excess water. In SW, water loss is replaced by the fish drinking and the excess of monovalent ions are excreted by the gills and divalent ions by the kidney (Evans, 1982; Kirsch *et al.*, 1981; Brown *et al.*, 1992). The co-ordination of these transporting activities is brought about by the actions of a cocktail of hormones. This chapter will focus on one of these hormones, the neurohypophysial peptide arginine vasotocin (AVT).

1.2 *Neurohypophysial peptides*

Neurohypophysial peptides are found in all the vertebrate classes and are characterized by their nine amino acid structure with a di-sulphide bridge linking the two cystine at positions 1 and 6 to give a ring structure (Acher, 1996). Teleost fish possess

Osmoregulation and Drinking in Vertebrates, edited by N. Hazon & G. Flik.
© 2002 BIOS Scientific Publishers Ltd, Oxford

two neurohypophysial peptides, AVT and isotocin. Although isotocin differs only from AVT by the amino acids present at positions 4 and 8, it does not have the same spectrum of physiological actions that have been identified for AVT. Currently we know little about the mechanisms that isotocin is involved in regulating. AVT is stored and released from the pituitary, but the peptide is synthesized in the cell bodies of the magnocellular neurons in the pre-optic nucleus in the hypothalamus (Moons *et al.*, 1989). AVT is synthesized in the hypothalamus as part of a larger pre-cursor molecule, pro-vasotocin (proVT). ProVT has been cloned from a number of fish species (Figueroa *et al.*, 1989; Heierhorst *et al.*, 1989; Hyodo *et al.*, 1991; Warne *et al.*, 2000) and structurally it comprises a signal peptide, the neurohypophysial peptide, a neurophysin and an N-terminal region which has a high homology with copeptin found in higher vertebrates. ProVT is translocated along the axons to the pituitary for storage and in response to the appropriate stimulus the precursor is enzymatically cleaved and AVT released into the blood. The last part of the AVT system is the receptors in the target tissues, which when activated alters the function of target tissue which ultimately feeds back on synthesis and release of AVT (*Figure 1*).

2. *Response of the AVT system to salinity challenge*

One approach to study the role a hormone plays in the control of osmoregulation is to impose an osmotic challenge on fish and record the response. There is a great deal of variation in the ability of individual fish species to adapt and respond to an osmotic challenge. The majority of teleost fish species are stenohaline and have adopted either a FW or SW existence and only have the osmoregulatory mechanisms to cope with one or other environment. There are however, a small group of

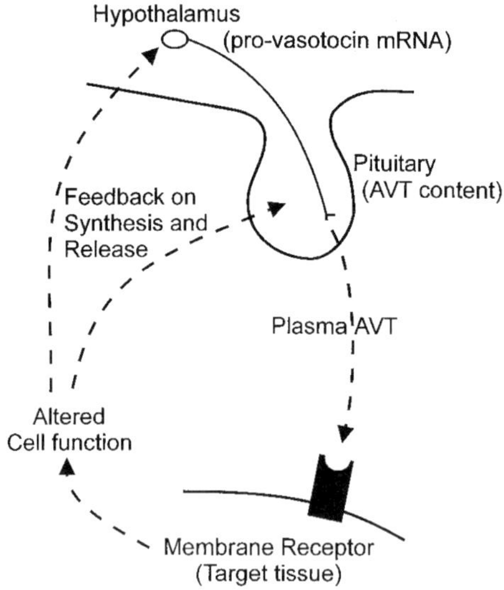

Figure 1. *Schematic diagram showing the feedback loop for the AVT system. AVT is synthesized as part of a larger precursor molecule in the magnocellular neurons in the hypothalamus. This precursor is stored in and AVT released into the circulation from the pituitary. Activation of receptors in the target tissue alters cell function which then feeds back on the levels of synthesis and release of AVT.*

euryhaline fish species that are capable of altering and controlling the function of transporting epithelia in osmoregulatory tissues enabling them to live in both FW and SW and move freely between them. This group of fish, in addition to being able to cope with the day to day perturbations that would be encountered in the relatively constant osmotic surrounding of SW or FW, also possess the physiological mechanisms to respond to and adapt to a new osmotic environment. The ability to adapt their physiology and survive in the extremes of both FW and SW makes them invaluable models in the study of osmoregulation in teleosts and they have been used extensively in examining the role AVT plays in osmoregulation.

2.1 *Secretion*

A number of studies have examined how the store of AVT in the pituitary and circulating AVT levels change in response to osmotic challenge. These studies fall into two basic categories, those that compare these parameters in fish chronically adapted to FW and SW and those that look at these parameters following acute challenge, when the fish are responding to a new challenge. It has been relatively straightforward to develop assays to quantify the large quantities of AVT stored in the pituitary. Once the AVT has been released into plasma, however, the circulating levels are in comparison very low. Until relatively recently reliable assays, which were sensitive enough to measure circulating levels, have not been available. The development of a sensitive radioimmunoassay (RIA) (Warne *et al.*, 1994) has rectified this problem and enabled the effects of osmotic challenge on circulating levels of AVT to be examined in the same way that the pituitary store of the peptide has been previously studied. Much of the work using the RIA for AVT has been in combination with experiments using the flounder (*Platichthys flesus*) as a model, although plasma AVT concentrations have also been measured in a number of other teleosts (Warne *et al.*, 2001).

2.2 *Chronic osmotic challenge*

When the pituitary store of AVT is compared in groups of euryhaline fish that have been adapted to FW and SW there is a lack of consistency in the results making interpretation difficult. In the European eel (*Anguilla anguilla*) the AVT store of the pituitary is higher in fish adapted to SW than those adapted to FW (Holder, 1968; Holder, 1970). In the flounder, pituitary levels of AVT in FW-adapted fish have been reported to be higher than those in SW (Perrott *et al.*, 1993), or there is no difference in the AVT store in FW- and SW-adapted fish (Bond *et al.*, 2001). In both the rainbow trout (*Oncorhynchus mykiss*) and medaka (*Oryzias latipes*) the pituitary stores of AVT in FW- and SW-adapted animals are not significantly different (Carlson and Holmes, 1962; Haruta *et al.*, 1991). Each of these studies has quantified AVT in different ways and this could account for some of the variability, also the length of time that fish are adapted to either FW or SW before sampling could also complicate the data. Another possibility may be that in some species of fish there is a difference in the pituitary store of AVT in SW and FW.

Circulating AVT levels measured by RIA in chronically adapted FW and SW species of euryhaline fish can be seen in *Figure 2*. No difference in the circulating levels of AVT was apparent between the FW- and SW-adapted fish. It appears that despite the contrasting osmotic demands of FW and SW, circulating levels of the

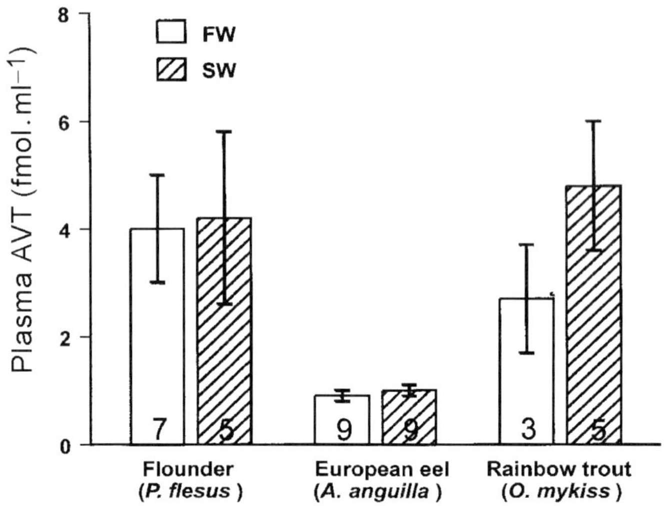

Figure 2. *Plasma AVT concentrations in three euryhaline teleost fish species adapted to FW and SW.* n *Values for each group are shown at base of columns.*

hormone in chronically adapted fish are not different, mirroring to some extent the absence of any consistent change in pituitary content under these conditions.

2.3 *Acute osmotic challenge*

The effect on pituitary store of AVT of an acute osmotic challenge, when the fish is responding to a new osmotic environment, is much clearer. In the flounder, rainbow trout and medaka the hyper-osmotic challenge of transfer from FW to SW results in a fall in the pituitary store of AVT (Carlson and Holmes, 1962; Haruta *et al.*, 1991; Perrott *et al.*, 1993). This response suggests that there has been a release of AVT from the pituitary in response to this challenge. The effect on the pituitary store of AVT of the reverse challenge of SW to FW transfer has been less well studied and there is only data from the flounder. Pituitary AVT was reported to show a slight increase following transfer of flounder from SW to FW although the rise was not statistically significant (Bond *et al.*, 2001).

Similarly, when fish are exposed to an acute osmotic challenge changes in circulating levels of AVT are seen. In the eel following transfer from FW to SW plasma AVT concentrations are increased at four days after transfer before returning to pre transfer levels (Balment *et al.*, 1993). Unpublished data we have from the flounder also shows that there is an increase in circulating AVT levels in response to transfer from FW to SW. When flounder are subject to the hypo-osmotic challenge of SW to FW transfer, there is a corresponding fall in the circulating levels of the peptide (Bond *et al.*, 2001).

When a fish is subject to an acute osmotic challenge there is a change in the secretion of AVT. Hyper-osmotic challenge causes the pituitary store of AVT to fall together with an increase in circulating levels of the peptide, indicating that there has been an increase in the release of peptide in response to this challenge. The increase in secretion following acute hyper-osmotic challenge is also supported by the observation that vasotocin neurons in the pro-optic nucleus of the goldfish (*Carassius auratus*) are stimulated when fish are exposed to salt solutions, suggesting release is occurring (Jasinski *et al.*, 1967). This response suggests that AVT is important when fish adapt to SW and that AVT regulates mechanisms that facilitate this adaptation. A hypo-osmotic challenge, although not having a great effect on the pituitary store, does result in a fall in circulating AVT levels suggesting that there is a reduced secretion. The response of the AVT system following this challenge indicates that AVT may not be important in the process of adaptation to FW.

2.4 *Synthesis*

Although proVT has been cloned from a number of teleost fish, few studies have examined how osmotic change affects mRNA expression. In rainbow trout proVT mRNA levels fell when fish were transferred to 80% SW from FW (Hyodo and Urano, 1991) which seems to imply that synthesis is reduced. In view of the changes in plasma and pituitary AVT in response to hyper-osmotic challenge, this result is difficult to reconcile with the hypothesis that AVT is important for adaptation to SW. If mRNA levels are indicative of levels of translation we might expect following the release of AVT in response to hyper-osmotic challenge, then proVT mRNA levels would increase rather than fall, as seen in this study, to help replace the depleted store of AVT in the pituitary.

2.5 *Stimuli for secretion*

When fish move between FW and SW or encounter other osmotic challenges, these are likely to result in changes in plasma osmolality and there may also be changes in blood volume. These parameters may therefore act as stimuli to alter AVT secretion. In a series of studies using chronically cannulated flounder, blood volume and plasma osmolality were manipulated and circulating AVT levels then measured. When a fish is subject to a hyper-osmotic challenge it would be predicted that the change in osmotic gradients would possibly result in a decrease in blood volume and an increase in plasma osmolality. Decreasing blood volume by haemorrhage, produced no change in plasma AVT concentration in the 60 minutes following this manoeuvre, suggesting that volume changes may not be an important trigger for AVT secretion (Warne and Balment, 1995). Raising plasma osmolality by intra-peritoneal injection of hypertonic NaCl did, however, result in a significant rise in circulating AVT levels, indicating that secretion was sensitive to changes in plasma osmolality or some component of it (Warne and Balment, 1995). Warne and Balment (1995) also reported a positive linear relationship between plasma AVT concentration and plasma osmolality over the normal ranges of these parameters in flounder held in SW. This relationship is also found between plasma AVT and plasma Na and Cl concentrations, the correlation between plasma AVT and Na for SW adapted flounder shown in *Figure 3*. The positive linear relationship between plasma AVT and osmolality and the increase in plasma AVT observed when plasma osmolality is raised, does suggest that these parameters are linked and that secretion of AVT is at least in part regulated by plasma osmolality in fish such as the flounder.

The experiments describing the responses of plasma and pituitary AVT levels to osmotic challenge presents somewhat of a paradox. There is an increase in the release of AVT in response to acute hyper-osmotic challenge suggesting that the

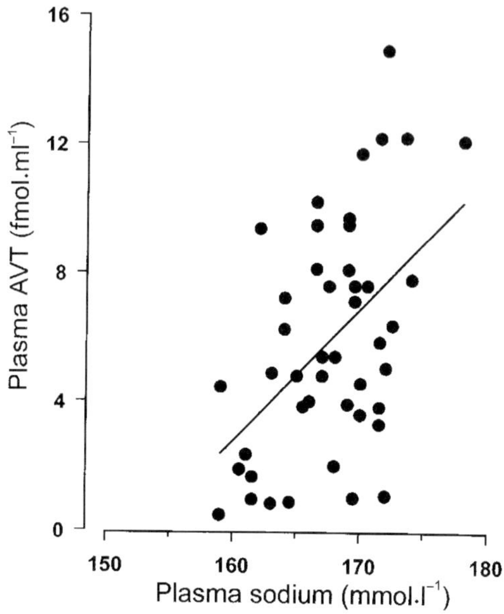

Figure 3. *Plasma AVT concentrations and plasma sodium are plotted for individual SW-adapted flounder. There is a positive correlation between these parameters ($r = 0.49$, $P < 0.001$).*

adaptive physiological mechanisms AVT is controlling are being up-regulated as part of the strategy for adaptation and survival. However, there is no real difference in plasma and pituitary levels of AVT in chronically adapted fish despite the evidence that the AVT system is being up-regulated for survival in SW. The profile of the change in plasma AVT levels following acute challenge suggests that there is a two-phase response in the adaptation to a new salinity. Initially there is a rise in plasma AVT following hyper-osmotic challenge (Balment *et al.*, 1993) and a fall following hypo-osmotic challenge (Bond *et al.*, 2001) but these changes are then followed by a return to pre-challenge values within a week of the original transfer. The initial change in circulating AVT levels indicates that in response to the acute osmotic challenge the rates of secretion are altered. The return of circulating AVT levels to control levels then suggests that the rates of secretion have returned to the levels seen in chronically adapted fish, suggesting that changes may be occurring during the acclimation process which are perhaps altering the actions and the responsiveness of target tissues to AVT. With the adaptation to a new osmotic environment there may be changes in the distribution and sensitivity of the AVT receptors, which are affecting the responsiveness of the target tissues to AVT.

3. Actions of AVT at target tissues

3.1 *Receptors*

AVT receptors have been cloned from two fish, a FW species, the white sucker, *Catostomus commersoni* (Mahlmann *et al.*, 1994) and also from the euryhaline flounder (Warne, 2001). The teleost AVT receptors are part of the super-family of G-protein-coupled receptors, with a characteristic seven trans-membrane spanning structure. Expression of the receptors in *Xenopus laevis* oocytes in conjunction with the technique of two electrode voltage clamping (*Figure 4*) has demonstrated that the receptors link to the phospholipase C-phosphatidylinositol (PLC-IP3) intra-cellular signalling pathway (Warne, 2001).

In this system the application of AVT to oocytes expressing the receptor, through the activation of the PLC-IP3 pathway, increases intra-cellular calcium which in turn opens calcium sensitive chloride channels resulting in an inward current that can be measured using a two electrode voltage clamp (*Figure 4*). The response to AVT can be blocked by injecting oocytes which express the receptor with inhibitors of the signalling pathway (Warne, 2001). The AVT receptors identified from fish have been shown to be specific to vasotocin (*Figure 4*). Isotocin, the second neurohypophysial peptide present in teleosts, is much less potent than AVT, with the receptor selectivity thought to be mediated by the sixth trans-membrane region and the third extra-cellular loop (Hausmann *et al.*, 1996).

Comparatively, the AVT receptors which have been identified from fish have many similarities with mammalian V1-type arginine vasopressin (AVP) receptors in terms of their sequence similarity and the linkage to intra-cellular signalling pathways. Neurohypophysial peptide receptors in mammals have been studied in greater detail than those in fish, and in terms of the intra-cellular signalling system associated with the receptor, two main classes of receptor have been identified in mammals. In addition to the PLC-IP3-linked V1-type receptor there is also a second class, termed a V2-type receptor, which through the activation of adenylate cyclase alters intra-cellular cyclic adenosine monophosphate (cAMP) levels (*Figure 5*). Although no vasotocin receptor has been identified in teleosts that links to adenylate cyclase and uses cAMP

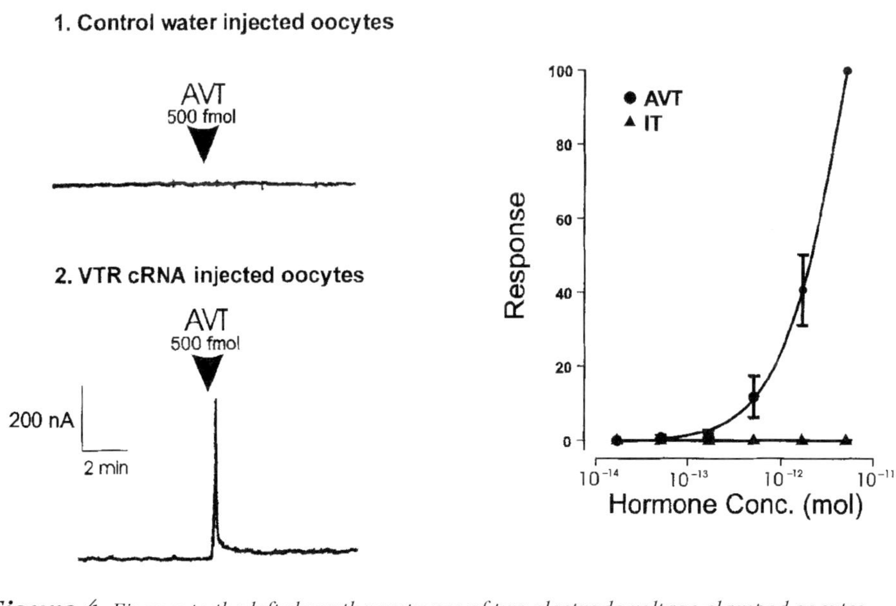

Figure 4. *Figures to the left show the response of two electrode voltage clamped oocytes which are expressing the flounder AVT receptor (2) or those that were control water injected to AVT (1). The figure to the right shows the dose–response relationship of the oocytes expressing the flounder AVT receptor to AVT and isotocin.*

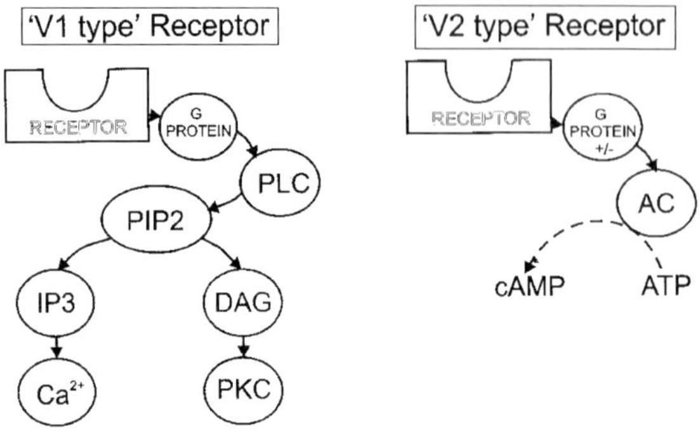

Figure 5. *Arginine vasopressin V1 and V2 signalling pathways. PLC, phospholipase C; PIP2, phosphatidyl-4,5-bisphosphate; IP3, inosotol-1,4,5-triphosphate; DAG, diacylglycerol; PKC, protein kinase C; AC, adenylate cyclase.*

as a second messenger, there are studies which suggest that such a receptor class may be present in teleost fish. These studies have focused on two elements in the signalling pathway downstream of the receptor and shown changes in their activity in response to AVT. One of these elements has been adenylate cyclase. Using membrane vesicles prepared from the gills of the rainbow trout, AVT was shown to cause a reduction in the activity of adenylate cyclase (Guibbolini and Lahlou, 1987), an effect mediated by an inhibitory G-protein (Guibbolini and Lahlou, 1992). This response is different to that seen in mammals when the V2-type receptor is activated by AVP. In mammals, activation of the V2 receptor increases intra-cellular cAMP (Wargent *et al.*, 1999), an action mediated by an increase in the activity of adenylate cyclase. Thus AVT is able to alter adenylate cyclase activity in fish but not in a way that would perhaps be anticipated. The other part of this signalling pathway that appears sensitive to AVT is the intra-cellular levels of cAMP. In isolated kidney tubules prepared from the rainbow trout, AVT has been shown to stimulate intra-cellular cAMP production in a dose-dependent manner (Perrott *et al.*, 1993). A response that is consistent with the effects of AVP on V2-type receptors in mammals.

Receptors in the target tissues mediate the actions of AVT and receptors for AVT appear to be present in the gills and kidney of teleosts, but it is not clear if receptors are present in the gastro-intestinal tract or whether AVT may have a regulatory action in this tissue.

3.2 *Gills*

The gills are an integral part of the mechanisms that control water and ion balance, being the site of active monovalent ion secretion in SW and uptake in FW. AVT receptors are present in the gill. mRNA for the PLC-IP3-linked vasotocin receptor is expressed in the gills (Mahlmann *et al.*, 1994; Warne *et al.*, 2001) and binding studies with iodinated vasotocin have demonstrated the presence of a single population of receptors in intact gill cells from the eel (Guibbolini *et al.*, 1988). Although receptors for AVT are present in the gills, the specific mechanisms which are regulated by AVT in relation to osmoregulation have yet to be established. Binding studies have demonstrated that receptor number is salinity-sensitive, the number of binding sites present in the gill cells of SW fish are higher than those of FW adapted animals (Guibbolini *et al.*, 1988). The inhibitory response of AVT on adenylate cyclase activity described by Guibbolini and Lahlou (1987, 1990), is also salinity-sensitive with fish adapted to SW showing a much greater response than those adapted to FW. These studies suggest that there is some change in the mechanisms that AVT is regulating in response to changes in salinity but what regulatory effect it has and how this then contributes to osmoregulation is unknown. AVT has been shown to have an effect on blood flow through the gills (Bennett and Rankin, 1986) and one possibility is that AVT may affect the blood flow through the gill altering blood distribution to different vascular beds. Another alternative is that AVT could have a direct effect on ion or water transport. Work by Maetz *et al.* (1964) did suggest that AVT could affect branchial ion transport, although results from this study were complicated by the associated renal effects of AVT making it difficult to determine if AVT was clearly affecting branchial ion transport.

3.3 *Kidney*

The teleost kidney plays very different roles in FW and SW. In FW it is responsible for eliminating water and in SW its primary osmoregulatory role is the elimination of divalent ions. Such is the difference in role of the teleost kidney in FW and SW that there are substantial differences in kidney structure between FW and SW fish and in euryhaline fish the kidney undergoes substantial structural changes when a fish moves and adapts to a new salinity (Hickman and Trump, 1969; Gray and Brown, 1987).

AVT receptor mRNA is expressed in the kidney of flounder (Warne, 2001) although in the FW species *C. commersoni* AVT receptor mRNA was reported to be absent (Mahlmann *et al.*, 1994). The renal actions of AVT suggest that functional receptors are present, and the effects of AVT on kidney function have been investigated using a number of techniques.

Chronically cannulated FW eels have been utilized as a model to study the effects of AVT on kidney function using renal clearance techniques (Chester Jones *et al.*, 1969; Henderson and Wales 1974; Babiker and Rankin, 1978). These studies have reported that AVT produces a diuresis following injection of high doses of AVT which were also accompanied by a systemic pressor response, while an antidiuresis followed injection of low AVT doses. The circulating levels that result from the injection of the various doses of AVT used in the studies are a key factor in interpreting these results. This information enables the responses to AVT to be evaluated in relation to whether the circulating levels of AVT produced after injection were physiologically relevant or pharmacological. In these earlier studies sufficiently sensitive assays for AVT were not available and it was not possible to measure circulating AVT concentrations. With the development of a sensitive RIA for AVT (Warne *et al.*, 1994) it has been possible to gain an idea of the circulating levels induced following the injection of various doses of AVT. Using chronically cannulated flounder, plasma samples were collected 20 minutes after the injection of AVT and resulting plasma AVT concentrations measured; these data are shown in *Table 1*.

Plasma AVT concentrations in teleost fish fall normally within the range 1–20 fmol. ml^{-1} (Warne *et al.*, 1994) and in the flounder this is more usually 1–10 fmol. ml^{-1}. It can be seen in *Table 1* that the AVT concentrations that are associated with a

Table 1. *Plasma AVT concentrations and presence or absence of an associated pressor response in flounder (*P. flesus*) 20 minutes after bolus injection of AVT.*

AVT dose injected/ per 100 g body weight	Plasma AVT concentration 20 minutes after injection (fmol ml^{-1})	Pressor response
500 pg	7.9 ± 1.4 (9)	No
5 ng	24.0 ± 5.6 (13)	Yes
50 ng	586.2 ± 129.0 (9)	Yes
500 ng	9518 ± 3654 (4)	Yes

Values are mean ± SEM, *n* values given in parenthesis.

systemic pressor response are significantly higher than normal circulating levels (Warne and Balment, 1997). With this information it is possible to re-evaluate earlier studies, and it would appear that the pressor response and diuresis following AVT injection (Chester Jones *et al.*, 1969; Henderson and Wales, 1974; Babiker and Rankin, 1978) is likely to have occurred in these studies at circulating levels of AVT which are pharmacological. The teleost kidney is unable to regulate glomerular filtration rates (GFR) independently of substantial changes in systemic blood pressure. In these studies it would appear that the diuresis has been a secondary result of AVT increasing systemic blood pressure. The antidiuretic action of AVT also reported in these studies is likely to be occurring at physiological levels of the peptide and probably represents the physiologically relevant effect of AVT in fish.

The antidiuretic effect of AVT in teleosts has also been demonstrated in more recent studies using an *in situ* perfused rainbow trout trunk preparation (Amer and Brown, 1995). By maintaining this model under constant perfusion pressure any complications caused by changes in systemic blood pressure were negated. Under these conditions AVT resulted in an antidiuresis even at doses which would, if used *in vivo*, result in a pressor response.

The transport maxima for glucose, which is a marker for the number of nephrons filtering, has also been measured in fish following the injection of AVT. In association with the fall in GFR following AVT injection, there was also a drop in the transport maxima for glucose (Henderson and Wales, 1974; Amer and Brown, 1995). It appears that the mechanisms by which AVT is altering urine production and causing an antidiuresis is through 'glomerular intermittency', the switching on or off of individual tubules. This phenomenon of glomerular intermittency has been confirmed by visual inspection of the tubules from rainbow trout following AVT treatment (Amer and Brown, 1995). The details of how this mechanism works have yet to be elucidated but the most likely explanation is that there are AVT receptors on the glomerular arterioles which regulate vascular smooth muscle contraction and glomerular blood flow. The AVP V1-type receptor is associated with vascular effects of AVP (the mammalian homologue of AVT) in mammals and the mRNA expression for the teleost homologue of this receptor has been detected in the kidney of the flounder (Warne, 2001).

The glomerular antidiuresis seen in fish is in contrast to the antidiuresis in mammals where AVP produces an antidiuretic effect by altering the permeability of the kidney tubules to water which is mediated by the V2 receptor class. The V2 receptor class has cAMP as its intra-cellular second messenger and as suggested earlier some data suggest that such a class of receptors for AVT may also be present in the teleosts (Perrott *et al.*, 1993). From mammalian studies we would predict that if teleosts possessed such a receptor it would regulate a mechanism for tubular modification of urine. Amer and Brown (1995) reported that AVT was able to alter tubular water reabsorption, although this was decreased at low doses of AVT and increased at high doses, making the interpretation difficult. If there are tubular AVT receptors in teleosts it is not clear how they would contribute to the overall role that AVT plays in regulating kidney function in fish.

4. Summary

The AVT system is a key component of the complex mechanism responsible for co-ordinating the response in teleost fish to hyper-osmotic challenge. The pituitary AVT store falls following transfer to SW from FW and this is accompanied by an increase in circulating AVT levels. This indicates that there has been release of AVT in response to this stimulus and the system is being up-regulated. The effect of an increase in secretion would appear to be predominantly to reduce GFR and urine production as is seen when AVT is injected into FW-adapted fish. This action is consistent with the requirements of hyper-osmotic challenge when water conservation becomes essential.

References

Acher, R. (1996) Molecular evolution of fish neurohypophysial hormones: Neutral and selective evolutionary mechanisms. *Gen. Comp. Endocrinol.* **102:** 157–172.

Amer, S. and Brown, J.A. (1995) Glomerular actions of arginine vasotocin in the *in situ* perfused trout kidney. *Am. J. Physiol.* **269:** R775–R780.

Babiker, M.M. and Rankin, J.C. (1978) Neurohypophysial hormone control of kidney function in the European eel (*Anguilla anguilla*) adapted to sea water or fresh water. *J. Endocrinol.* **78:** 347–58.

Balment, R.J., Warne, J.M., Tierney, M. and Hazon, N. (1993) Arginine vasotocin and fish osmoregulation. *Fish Physiol. Biochem.* **11:** 189–194.

Bennett, M.B. and Rankin, J.C. (1986) The effect of neurohypophysial hormones on the vascular resistance of the isolated perfused gill of the European eel *Anguilla anguilla* L. *Gen. Comp. Endocrinol.* **64:** 60–66.

Bond, H., Winter, M., Warne, J.M. and Balment, R.J. (2002) Plasma concentrations of arginine vasotocin and urotensin II are reduced following transfer of flounder (*Platichthys flesus*) from seawater to fresh water. *Gen. Comp. Endocrinol.* **125:** 113–120.

Brown, J.A., Rankin, J.C. and Yokota, S.D. (1992) Glomerular haemodynamics and filtration in single nephrons of non-mammalian vertebrates. In: *New Insights in Vertebrate Kidney Function* (eds J.A. Brown, R.J. Balment and J.C. Rankin). Cambridge University Press, Cambridge, pp. 1–44.

Carlson, I.H. and Holmes, W.N. (1962) Changes in the hormone content of the hypothalamo-hypophysial system of the rainbow trout (*Salmo gairdneri*). *J. Endocrinol.* **24:** 23–32.

Chester Jones, I., Chan, D.K.O. and Rankin, J.C. (1969) Renal function in the European eel (*Anguilla anguilla* L.): effects of the caudal neurosecretory system, corpuscles of Stannius, neurohypophysial peptides and vasoactive substances. *J. Endocrinol.* **43:** 21–31.

Evans, D.H. (1982) Salt and water exchange across vertebrate gills. In: *Gills* (eds D.F. Houlihan, J.C. Rankin and T.J. Shuttleworth). Cambridge University Press, Cambridge, pp. 149–171.

Figueroa, J., Morley, S.D., Heierhorst, J., Krentler, C., Lederis, K. and Richter, D. (1989) Two isotocin genes are present in the white sucker *Catostomus commersoni* both lacking introns in their protein coding regions. *EMBO J.* **8:** 2873–2877.

Gray, C.J. and Brown, J.A. (1987) Glomerular ultrastructure of the trout, *Salmo gairdneri*: effects of angiotensin II and adaptation to seawater. *Cell Tis. Res.* **249:** 437–442.

Guibbolini, M.E. and Lahlou, B. (1987) Neurohypophysial peptide inhibition of adenylate cyclase activity in fish gills. *FEBS Letters* **220:** 98–102.

Guibbolini, M.E. and Lahlou, B. (1990) Evidence for the presence of a new type of neurohypophysial hormone receptor in fish gill epithelium. *Am. J. Physiol.* **258:** R3–R9.

Guibbolini, M.E. and Lahlou, B. (1992) G_i protein mediates adenylate cyclase inhibition by neurohypophyseal hormones in fish gill. *Peptides* **13:** 865–871.

Guibbolini, M.E., Henderson, I.W., Mosley, W. and Lahlou, B. (1988) Arginine vasotocin binding to isolated branchial cells of the eel: effect of salinity. *J. Mol. Endocrinol.* **1:** 125–130.

Haruta, K., Yamashita, T. and Hawashima, S. (1991) Changes in the arginine vasotocin content of the pituitary of the medaka (*Oryzias latipies*) during osmotic stress. *Gen. Comp. Endocrinol.* **83:** 327–336.

Hausmann, H., Richters, A., Kreienkamp, H-J., Meyerhof, W., Mattes, H., Lederis, K., Zwiers, H. and Richter, D. (1996) Mutational analysis and molecular modelling of the nonapeptide hormone binding domains of the [Arg^8] vasotocin receptor. *Proc. Natl Acad. Sci. USA* **93:** 6907–6912.

Heierhorst, J., Morley, S.D., Figueroa, J., Krentler, C., Lederis, K. and Richter, D. (1989) Vasotocin and isotocin precursors from the white sucker, *Catostomus commersoni*: Cloning and sequence analysis of the cDNAs. *Proc. Natl Acad. Sci. USA* **86:** 5242–5246.

Henderson, I.W. and Wales, N.A.M. (1974) Renal diuresis and antidiuresis after injections of arginine vasotocin in the fresh water eel (*Anguilla anguilla* L.). *J. Endocrinol.* **61:** 487–500.

Hickman, C.P. and Trump, B.F. (1969) The kidney. In: *Fish Physiology*, Vol. I (eds W.S.Hoar and D.J.Randall). Academic Press, New York, pp. 91–239.

Holder, F.C. (1968) Mise en évidence et dosage de l'activé de type ocytocique d'homogénats totaux et de fractions granulaires du systéme préoptico-hypophysaire de l'Anguille (*Anguilla anguilla* L.). *Gen. Comp. Endocrinol.* **11:** 235–242.

Holder, F.C. (1970) Séparation chromatographique et dosage de l'arginine-vasotocine et de l'isotocine de l'hypophyse et du noyau préoptique chez l'anguille de Mer du Nord. *Experientia* **26:** 1199–1200.

Holmes, W.N. and Donaldson, E.M. (1969) The body compartments and the distribution of electrolytes. In: *Fish Physiology*, Vol. I (eds W.S. Hoar and D.J. Randall). Academic Press, New York, pp. 1–89.

Hyodo, S. and Urano, A. (1991) Changes in expression of provasotocin and proisotocin genes during adaption to hyper- and hypo-osmotic environments in rainbow trout. *J. Comp. Physiol. B.* **161:** 549–556.

Hyodo, S., Kato, Y., Ono, M. and Urano, A. (1991) Cloning and sequence analysis of cDNAs encoding vasotocin and isotocin precursors of chum salmon, *Oncorhynchus keta*: evolutionary relationships of neurohypophysial hormone presursors. *J. Comp. Physiol. B.* **160:** 601–608.

Jasinski, A., Gorbman, A. and Hara, R.J. (1967) Activation of the preoptico-hypophysial neurosecretory system through olfactory afferents in fishes. In: *Neurosecretion* (ed. F. Strtinsky). Springer, New York, pp. 100–123.

Kirsch, R., Meens, R. and Meister, M. (1981) Osmoregulation chez les téléostéens marins: role des branchies et du tube digestif. *Bulletin Société Zoologique de France* **106:** 31–37.

Maetz, J., Bourguet, J., Lahlou, B. and Hourdry, J. (1964) Peptides neurohypophysiares et osmoregulation chez *Carassius auratus*. *Gen. Comp. Endocrinol.* **4:** 508–522.

Mahlmann, S., Meyerhof, W., Hausmann, H., Heierhorst, J., Schönrock, C., Zwiers, H., Lederis, K. and Richter, D. (1994) Structure, function, and phylogeny of [Arg^8] vasotocin receptors in teleost fish and toad. *Proc. Natl Acad. Sci. USA* **91:** 1342–1345.

Moons, L., Cambre, M., Batten, T.F.C. and Vandesande, F. (1989) Autoradiographic localisation of binding sites for vasotocin in the brain and pituitary of the sea bass (*Dicentrarchus labrax*). *Neurosci. Lett.* **100:** 11–16.

Perrott, M.N., Sainsbury, R.J. and Balment, R.J. (1993) Peptide hormone stimulated second messenger production in the teleostean nephron. *Gen. Comp. Endocrinol.* **89:** 387–395.

Wargent, E.T., Burgess, W.J., Laycock, J.F. and Balment, R.J. (1999) Separate receptors mediate oxyticon and vasopressin stimulation of cAMP in rat inner medullary collecting duct cells. *Exp. Physiol.* **84:** 17–25.

Warne, J.M. (2001) Cloning and characterisation of an arginine vasotocin receptor from the euryhaline flounder *Platichthys flesus*. *Gen. Comp. Endocrinol.* **122:** 312–319.

Warne, J.M. and Balment, R.J. (1995) Effect of acute manipulation of blood volume and osmolality on plasma [AVT] in seawater flounder. *Am. J. Physiol.* **269:** R1107–R1112.

Warne, J.M. and Balment, R.J. (1997) Changes in plasma arginine vasotocin (AVT) concentration and dorsal aortic blood pressure following AVT injection in the teleost, *Platichthys flesus*. *Gen. Comp. Endocrinol.* **105:** 358–364.

Warne, J.M., Hazon, N., Rankin, J.C. and Balment, R.J. (1994) A radioimmunoassay for the determination of arginine vasotocin (AVT): Plasma and pituitary concentrations in fresh- and seawater fish. *Gen. Comp. Endocrinol.* **96:** 438–444.

Warne, J.M., Hyodo, S., Harding, K.E. and Balment, R.J. (2000) Cloning of pro-vasotocin and pro-isotocin cDNA's from the flounder *Platichthys flesus*: levels of hypothalamic mRNA following acute osmotic challenge. *Gen. Comp. Endocrinol.* **119:** 77–84.

Warne, J.M., Harding, K.E. and Balment, R.J. (2002) Review: Neurohypophysial hormones and renal function in fish and mammals. *Comp. Biochem. Physiol.* (in press).

7

Facilitative effects of angiotensin II on prolactin cell responses to osmotic stimulation in tilapia

T. Hirano, T.A. Leedom, A.P. Seale and E.G. Grau

1. Introduction

Prolactin (PRL) has long been known to be an essential freshwater (FW) osmoregulatory hormone in many euryhaline teleosts, including the tilapia *Oreochromis mossambicus*. Reflecting its role in FW, PRL cell activity is higher in fish acclimated to FW than in those acclimated to seawater (SW) (Hirano, 1986; Brown and Brown, 1987; McCormick, 2001). It has also been well established that PRL release from the tilapia pituitary *in vitro* increases as medium osmotic pressure is reduced (Nishioka *et al.*, 1988; Grau and Helms, 1990; Grau *et al.*, 1994). Recently, Shepherd *et al.* (1999) have clearly shown that changes in plasma osmolality *in vivo* also exert a direct regulatory action on PRL release and gene expression in the tilapia pituitary gland. In contrast to the direct effect of extra-cellular osmolality on PRL release, however, the hypothalamic control of PRL release in tilapia is not fully understood. As in many vertebrate species, PRL secretion in tilapia seems to be under strong inhibitory control by somatostatin (Grau *et al.*, 1982; Grau *et al.*, 1985; Nishioka *et al.*, 1988; Helms *et al.*, 1991). On the other hand, hypothalamic factors that have been shown to stimulate PRL release in tilapia include three native forms of gonadotropin-releasing hormones (Weber *et al.*, 1997) and thyrotropin-releasing hormone (Barry and Grau, 1985), the latter being effective only after pre-treatment with estradiol. Recently, we have found that *Carassius* RFamide (C-RFa), a fish homologue of mammalian PRL-releasing peptide (PrRP), also exhibits PRL-releasing activity in tilapia (Seale *et al.*, 2002). However, it was necessary to pre-incubate the pituitary gland in hyper-osmotic medium for more than 18 h to see the stimulatory effect of C-RFa; C-RFa had no effect without pre-incubation. The effect of mammalian PrRP was equivocal. The difficulty in producing a stimulatory effect of

Osmoregulation and Drinking in Vertebrates, edited by N. Hazon & G. Flik.
© 2002 BIOS Scientific Publishers Ltd, Oxford

hypothalamic peptides on PRL release from the tilapia pituitary *in vitro* may be due to the high baseline release of PRL in the absence of strong inhibitory control by the hypothalamus (Seale *et al.*, 2002).

In mammals, the entire renin–angiotensin system (RAS) is present in the hypothalamus as well as in the pituitary gland (Saavedra, 1992). It has also been shown that angiotensin II (ANG II) stimulates PRL release through the AT1 receptor through a Ca^{2+}-dependent process (Malarkey *et al.*, 1987; Diaz-Torga *et al.*, 1998; Iglesias *et al.*, 2000). While we were conducting experiments on plasma clearance of bovine growth hormone in tilapia, we became aware that repeated blood withdrawal caused a reduction in plasma osmolality, accompanied by a marked increase in plasma PRL levels (more than 400 ng ml^{-1}). Since hypo-volaemia caused by hemorrhage is known to stimulate RAS and drinking not only in mammals but also in fish (Kobayashi and Takei, 1996; Fitzsimons, 1998), we hypothesized that the marked increase in plasma PRL after blood withdrawal may be due to a synergistic effect of ANG II and reduced osmolality. This chapter describes our recent experiments indicating a stimulatory effect of ANG II on PRL secretion in the tilapia. Possible mechanisms that may facilitate effects of ANG II on responses of tilapia PRL cell to osmotic stimulation are also discussed.

2. Effects of blood withdrawal on plasma PRL levels

2.1 *Effects of repeated blood withdrawal*

Euryhaline tilapia (*Oreochromis mossambicus*) were obtained from a population maintained at the Hawaii Institute of Marine Biology, University of Hawaii. They were kept in either 60- or 600-litre tanks in FW. They were fed twice daily with Pro Form (Agro Pacific, Chilliwaeck, BC, Canada), approximately 5% of body weight per day. Water temperature was maintained at 26 ± 2°C.

Blood was removed serially from fish weighing 100–150 g, at 0, 1, 2, 5 and 20 h. Each sample consisted of 2–3% of total blood volume, estimated as 7% of the body weight (Okimoto *et al.*, 1994). Blood was taken from the caudal vessels *via* a heparinized syringe (200 U ml^{-1} ammonium heparin). Plasma was separated by centrifugation (5 min at 10 000 *g*). The tilapia pituitary releases two distinct PRL molecules, PRL_{177} and PRL_{188}, which are encoded by separate genes (Specker *et al.*, 1985; Yamaguchi *et al.*, 1988; Rentier-Delrue *et al.*, 1989). Concentrations of the two PRLs and growth hormone (GH) in the plasma were determined using homologous radioimmunoassays (RIAs) (Ayson *et al.*, 1993). Plasma osmolality was determined using a vapour pressure osmometer (Wescor 5100C, Logan, UT). All values represent means ± standard errors of the mean (SEM). Differences were determined by one-way or two-way analysis of variance (ANOVA) followed by Duncan's multiple-range test or by Mann–Whitney *U*-test. Calculations were performed using a computer program, STATISTICA (StatSoft, Tulsa, OK).

As shown in *Figure 1*, the plasma level of PRL_{188} at the start of experiment was around 5 ng ml^{-1}. It increased slightly but significantly ($P < 0.05$) after 1 h, increased further after 2 h, and reached a peak level (more than 100 ng ml^{-1}) after 5 h or at the time of the 4th blood withdrawal. The high PRL level was still maintained after 20 h. Plasma levels of PRL_{177} also increased markedly after blood withdrawal (data not shown). Although a slight and significant increase occurred in plasma GH after 5 h,

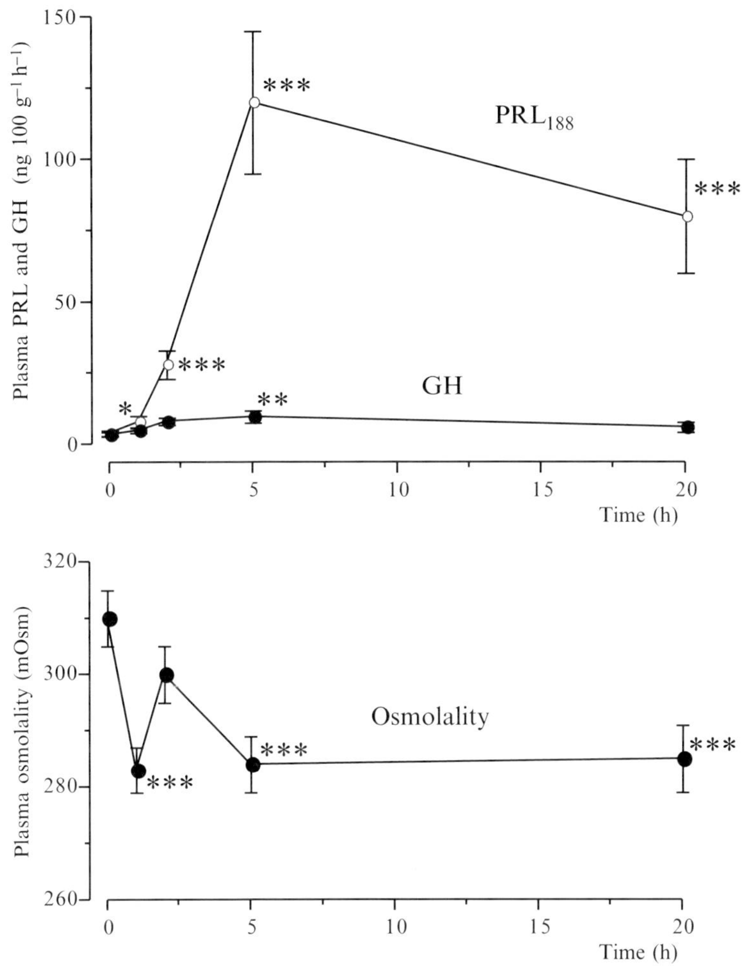

Figure 1. *Effects of repeated blood withdrawal on plasma levels of PRL1$_{88}$, GH and osmolality in tilapia (unpublished data from Seale, Hirano and Grau). Blood was removed serially from FW fish (weighing 100–150 g) at 0, 1, 2, 5 and 20 h. Each sample consisted of 2–3% of total blood volume. Vertical bars indicate mean ± SEM (n = 6–8). **, *** Significantly different from the control (time 0) at* P *< 0.01 and* P *< 0.001, respectively.*

the pattern of change in plasma GH was completely different from that of the PRLs, suggesting that the increased hormone levels after blood withdrawal are specific to PRLs.

The increase in plasma PRL seems to be correlated with the changes in plasma osmolality. The initial level of plasma osmolality was about 310 mOsm. Blood withdrawal caused a significant ($P < 0.001$) reduction in plasma osmolality after 1 h, which rebounded at 2 h, but decreased again after 5 h. The low level (about 280 mOsm) was still maintained after 20 h (*Figure 1*). Repeated blood withdrawal in the tilapia is always accompanied by significant decreases in haematocrit values (Leedom, unpublished observation). Reduction in plasma osmolality and haemodilution following haemorrhage have been reported in several teleost species (Nishimura *et al.*, 1979; Carroll *et al.*, 1984; Ogilvy *et al.*, 1988; Takei, 1988). Since the blood cells were not re-introduced in the present experiment, the proportion of blood cells in circulation should have remained stable. Therefore, the reduction in haematocrit observed in the present study is likely the result of restoration of blood volume by water from the other compartments of the body. Thus, tilapia seem to maintain blood volume at the expense of osmolality following blood removal.

Reflecting the important role of PRL in FW adaptation, plasma levels of PRLs in tilapia are inversely related to plasma osmolality. The movement of a FW-acclimated tilapia to SW produces a rapid, and sometimes large, increase in plasma osmolality. Conversely, the transfer of a fish in SW to FW elicits a rapid drop in blood osmolality, to the level about 280 mOsm (Yada *et al.*, 1994). The plasma of fish acclimated to FW exhibits significantly higher levels of PRLs (10–20 ng ml^{-1}) than the plasma of fish acclimated to SW (less than 5 ng ml^{-1}). Transfer from SW to FW elicits a marked increase in PRL levels within 6 h, the maximum increase observed being 40 ng ml^{-1} (Ayson *et al.*, 1993; Yada *et al.*, 1994). Therefore, it is unlikely that the increase in plasma PRL after repeated blood withdrawal to the level above 100 ng ml^{-1} is due only to the reduced plasma osmolality, since the degree of the change in plasma osmolality after blood removal was the same as that after transfer from SW to FW.

2.2 *Effects of single withdrawal of 20% of total blood volume*

In the next experiment, to confirm the effect of blood removal on plasma PRL levels, the effect of a single blood withdrawal (20% of estimated blood volume; approximately 500 μl) was examined in fish weighing 30–45 g. A minimal blood sample (< 50 μl) was taken from the control groups. Blood samples were also taken at the end of the experiment (after 5 h). Since hypo-volaemia caused by haemorrhage has been known to induce drinking not only in fish but also in vertebrates in general, drinking rate was also estimated in this experiment. Drinking rate was estimated following Kobayashi *et al.* (1983), as modified by Eckert *et al.* (2001). Immediately after blood withdrawal, fish were transferred to a 60-l glass aquaria containing phenol red solution (0.004%). After 5 h, the intestine was dissected, and its contents were washed with 3 ml distilled water into a Petri dish. The volume of ingested water was estimated from the amount of phenol red in the intestine.

As shown in *Table 1*, the plasma level of PRL_{188} of blood-depleted fish was significantly ($P < 0.001$) higher than the levels of the control fish. There was no significant difference in PRL_{177} due to the large individual variation in both control and experimental groups. Unlike the fish in the previous experiment, in which blood was taken repeatedly (cf. *Figure 1*), there was no difference in plasma osmolality. The drinking rate of the depleted fish was significantly ($P < 0.001$) greater than that of the control fish.

Stimulation of drinking after haemorrhage in tilapia is in accord with that reported for other species (Kobayashi and Takei, 1996; Fitzsimons, 1998). The absence of changes in plasma osmolality may be due to restoration of plasma ion levels during the experimental period (5 h), at least in part, by the increased PRL secretion. There is overwhelming experimental support for the role of PRL in promoting ion uptake not only in euryhaline fish but also in stenohaline fish in FW (Hirano, 1986; Brown and Brown, 1987; Eckert *et al.*, 2001; McCormick, 2001). In a separate experiment, a significant increase in gill Na^+,K^+-ATPase activity was observed after repeated blood

Table 1. *Effects of single withdrawal of 20% of total blood volume on plasma PRLs, plasma osmolality and drinking rate in tilapia acclimated to FW (unpublished data from Leedom, Hirano and Grau)*

	PRL_{188} (ng ml^{-1})	PRL_{177} (ng ml^{-1})	Plasma osmolality (mOsm)	Drinking rate (ml 100 g^{-1} h^{-1})
Control	14.1 ± 2.39[a]	49.1 ± 12.3	314 ± 4.2	0.06 ± 0.01
Blood removal	32.4 ± 1.18***	96.4 ± 28.6	319 ± 4.1	0.24 ± 0.01***

[a]Mean ± SEM (n = 6). ***Significantly different from the control at $P < 0.001$.

withdrawal in the tilapia in FW indicating activation of the ion uptake mechanism by the gills (Leedom, unpublished observation).

Haemorrhage is known to stimulate the RAS in all the vertebrates, and the dipsogenic action of ANG II has also been well established in fish (Tierney *et al.*, 1995; Kobayashi and Takei, 1996; Fuentes and Eddy, 1997; Tsuchida and Takei, 1998; Tsuchida and Takei, 1999). It is highly likely, therefore, that the reduced plasma osmolality after repeated blood withdrawal (*Figure 1*) may be a result, at least in part, of restoration of blood volume by ingesting environmental water, induced possibly by ANG II. When blood is taken from the tilapia, they may restore blood volume by drinking water from the environment at the expense of plasma osmolality. The fact that a single withdrawal of 20% of blood resulted in an increase in plasma PRL without affecting plasma osmolality suggests the involvement of other factors, such as ANG II, in the stimulation of PRL secretion after blood withdrawal.

3. Effects of ANG II on PRL secretion

3.1 In vivo *effect*

The effect of intra-peritoneal injection of ANG II on plasma PRLs and GH was examined using FW-adapted tilapia (20–40 g). They were injected with ANG II ([1]Asp, [5]Val-ANG II, Peptide Institute, Osaka, Japan) at a dose of 0.1 or 1.0 μg g^{-1}. The control group received saline (0.9% NaCl) injection. Blood samples were taken 1 h after the injection. As shown in *Table 2*, a significant ($P < 0.05$) increase in PRL_{177} was observed after injection of the low dose (0.1 μg g^{-1}) of ANG II. Plasma levels of both PRL_{188} and PRL_{177} were increased significantly ($P < 0.01$) after injection of the high dose (1.0 μg g^{-1}). On the other hand, the high dose of ANG II significantly reduced ($P < 0.01$) plasma GH. No significant change was seen in plasma osmolality.

In teleost fish, intra-peritoneal injections of ANG II induced drinking in euryhaline species such as eels, flounder and killifish (Takei *et al.*, 1979; Malvin, 1980; Carrick and Balment, 1983; Hirano and Hasegawa, 1984). In these experiments, large doses of ANG II (0.01–1 μg g^{-1}) were injected, since its plasma concentration increases only transiently because of its short life in plasma (Naden *et al.*, 1985). Kobayashi *et al.* (1983) reported that ANG II had no effect on drinking in tilapia (*Oreochromis mossambicus*), even at a dose as high as 1 μg g^{-1}. According to Tsuchida and Takei

Table 2. *Effects of intra-peritoneal injection of ANG II ([1]Asp, [5]Val-ANG II) on plasma levels of PRL_{188}, PRL_{177} and GH, and plasma osmolality (unpublished data from Leedom, Hirano and Grau)*

Treatment	Plasma PRL_{188} (ng ml^{-1})	Plasma PRL_{177} (ng ml^{-1})	Plasma GH (ng ml^{-1})	Plasma osmolality (mOsm)
Control (saline)	15.7 ± 5.6[a]	22.8 ± 3.1	2.8 ± 0.7	320 ± 8.5
Ang II (0.1 mg g^{-1})	20.0 ± 6.6	33.6 ± 2.2*	1.2 ± 0.2	334 ± 11.2
Ang II (1.0 mg g^{-1})	34.6 ± 8.2**	81.5 ± 14.6**	0.9 ± 0.1**	316 ± 4.8

[a]Mean ± SEM (n = 9). *,**Significantly different from the control at $P < 0.05$ and $P < 0.01$, respectively.

(1999), intra-arterial infusion of a physiological dose of ANG II (0.1 pmol kg^{-1} min^{-1} or 0.13 pg^{-1} min^{-1}) induced drinking in the eel in FW, without altering plasma ANG II concentration or arterial pressure. In the present study, large doses of ANG II were necessary to obtain the increase in plasma PRLs. To clarify physiological roles of ANG II in PRL secretion in tilapia, additional studies will be required by infusing small doses of the hormone through an intra-arterial or intravenous catheter, while monitoring its plasma level and blood pressure.

3.2 In vitro *effect*

The *in vitro* effect of ANG II on PRL secretion was examined using the pituitary glands isolated from FW-adapted fish (30–60 g). The pituitary glands were placed individually in a well of a 96-well culture plate containing 100 μl of isotonic (330 mOsmol kg^{-1}) Krebs bicarbonate-Ringer solution, containing glucose (500 mg l^{-1}), glutamine (290 mg l^{-1}) and Eagle's minimal essential medium as described by Wigham *et al.* (1977). Pituitary glands were pre-incubated for 18–20 h at 26–28°C on a gyratory platform (80 rpm) under a humidified atmosphere of 95% O_2 and 5% CO_2. After the pre-incubation, the pituitaries were incubated for 8 h with 0, 0.1, 1.0 or 10 nM ANG II ([1]Asp, [5]Val-ANG II). The medium was replaced at 1, 2 and 4 h. As shown in *Figure 2*, ANG II stimulated PRL release over time dose-dependently ($P < 0.05$). The effect was already apparent after 1 h of incubation and lasted for 8 h.

The RAS was originally recognized as a circulating hormonal system in mammals. Renin (produced by the juxtaglomerular cells in the kidney) and angiotensinogen (produced by the liver) are released into the general circulation, and ANG I is produced by renin in the circulation. ANG I is converted to ANG II as it circulates by angiotensin-converting enzyme (ACE). In addition to this circulating RAS, it is now clear in mammals that tissue-specific RAS occurs not only in the kidney but also in the hypothalamus and the pituitary with the potential for paracrine actions (Hackenthal *et al.*, 1990; Saavedra, 1992). Recently, Brown *et al.* (2000) reported the expression of the angiotensinogen gene in the trout kidney, providing the first molecular evidence for a tissue-specific RAS in fish. ANG II receptors have been identified in the gills, kidney, intestine and liver in several teleost species (Russell *et al.*, 2001). Further studies are called for to clarify the presence of RAS and ANG II receptor in the fish pituitary, specifically in PRL cells.

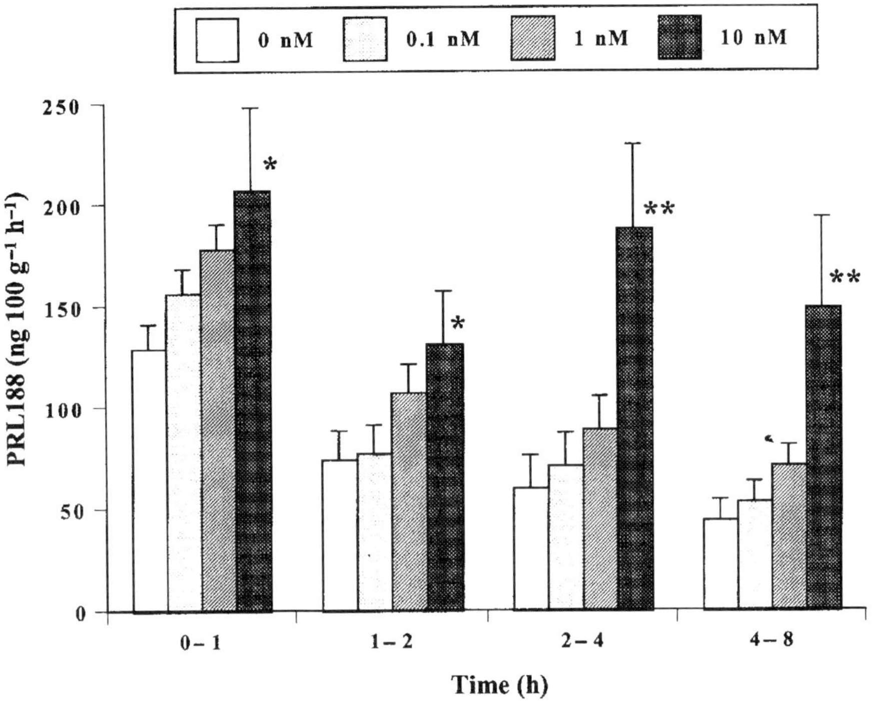

Figure 2. *Effects of ANG II on release of PRL_{188} from the tilapia pituitary gland* in vitro *(unpublished data from Eckert, Hirano and Grau). Pituitary glands were pre-incubated for 24 h, and then exposed to 0.1–10 nM ANG II ([1]Asp, [5]Val-ANG II). Incubation medium was changed at 1, 2, 4 and 8 h. PRL release is normalized to body weight and expressed as a rate during each time interval. Vertical bars indicate mean ± SEM* (n = 6–8). **, ** Significantly different from the control (0 nM) at each time point at* $P < 0.05$ *and* $P < 0.01$*, respectively.*

4. Facilitation of PRL cell responses to osmotic stimulation by ANG II

In the present study, repeated blood withdrawal or a single withdrawal of 20% of total blood from the tilapia caused a marked increase in plasma levels of PRLs. The increase was well correlated with a decrease in plasma osmolality. Single blood withdrawal (20% of total blood) stimulated drinking in the fish in FW. Furthermore, ANG II stimulated PRL secretion both *in vivo* and *in vitro*. It is likely, therefore, that the tilapia restores blood volume after blood withdrawal by drinking environmental water at the expense of plasma osmolality, and that ANG II is involved in both initiation of drinking and increased secretion of PRLs. The reduced plasma osmolality would stimulate PRL secretion by itself. There is overwhelming experimental support for the role of PRL in promoting ion uptake in fish in FW (Hirano, 1986; Brown and Brown, 1987). On the other hand, although cortisol has been identified largely as a SW-adapting hormone, primarily stimulating ion excretion by the gills, an increasing body of evidence indicates that cortisol is also involved in ion uptake in FW (McCormick, 2001). Thus, the reduced plasma osmolality after blood withdrawal would be restored by the ion-retaining action of PRL and cortisol in due course (*Figure 3*).

On the other hand, repeated blood withdrawal from the fish in SW caused a significant increase in plasma osmolality, and no increase was observed in plasma PRL levels. Indeed, PRL levels were lower than that of the fish in FW (Leedom, unpublished observation). This suggests that increased plasma osmolality after blood removal in the SW fish inhibits PRL secretion and overrides the stimulating effect of ANG II. As stated above, increases in extra-cellular osmolality attenuate PRL secretion from the tilapia pituitary gland both *in vivo* and *in vitro* (Grau and Helms, 1990; Grau *et al.*, 1994). However, ANG II is still effective in stimulating PRL secretion from the tilapia pituitary gland incubated in hypertonic medium (360 mOsm, Leedom, unpublished observation). Therefore, inhibitory factors such as somatostatin may be operating to inhibit PRL secretion in the fish in SW (Grau *et al.*, 1982; Grau *et al.*, 1985; Helms *et al.*, 1991). This is conceivable since the sodium-retaining action of PRL in teleosts is antagonistic to SW adaptation (Hirano, 1986; Brown and Brown, 1987). Restoration of elevated plasma osmolality in SW would be achieved by the action of cortisol, GH or other hormones (Eckert *et al.*, 2001; McCormick, 2001).

ANG II has also been known to exert excitatory effects on supraoptic neurosecretory neurons in the mammalian hypothalamus. Increases in extra-cellular fluid osmolality cause the neurosecretory cells to shrink and a depolarization of the cells through stretch-inactivated cation channels upon cell shrinking (Bourque, 1998). Recently, Chakfe and Bourque (2000) reported that excitatory effects of ANG II are due to the stimulation of the stretch-inactivated channels. As stated in the introduction of this chapter, PRL cells of tilapia are osmosensitive, responding to small changes in extra-cellular osmolality (Nishioka *et al.*, 1988; Grau *et al.*, 1994). A decrease in extra-cellular osmolality of less than 5% (325–310 mOsm) results in a marked elevation in PRL release both *in vivo* and *in vitro*. A decrease in extra-cellular osmolality will increase cell volume through passive influx of water. This seems to initiate an influx of extra-cellular Ca^{2+}, possibly through stretch-gated channels, and a

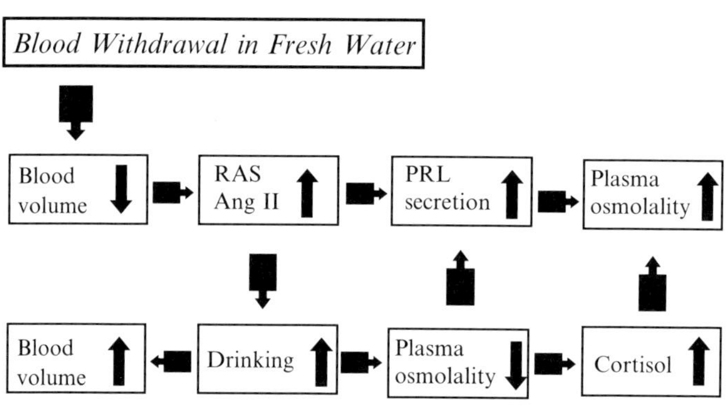

Figure 3. *Facilitation by ANG II of PRL cell responses to osmotic stimulation. Blood withdrawal (reduced blood volume) will stimulate RAS and production of ANG II, which in turn will stimulate drinking and PRL secretion at the same time. The recovery of blood volume, partly by ingested water, will result in the reduction in plasma osmolality, which will also stimulate PRL secretion. The reduced plasma osmolality would be restored by the ion-retaining action of PRL and cortisol in due course.*

consequent rise in intra-cellular Ca^{2+}, which mediates the increase in PRL release. In mammals, ANG II stimulates PRL release by interacting with the AT_1 receptor through a Ca^{2+}-dependent process (Malarkey *et al.*, 1987; Diaz-Torga *et al.*, 1998; Iglesias *et al.*, 2000). In the pituitary, stimulation of the AT_1 receptor causes phospholipase C-mediated hydrolysis of membrane phosphoinositides, leading to the formation of diacylglycerol and inositol phosphates and subsequently to the activation of protein kinase C and to an increase in intra-cellular Ca^{2+} levels (Canonico and MacLeod, 1986; Malarkey *et al.*, 1987). Chakfe and Bourque (2000) postulated that the activation of the pathways leading to channel phosphorylation or dephosphorylation may be involved in the modulatory effects of ANG II in stimulating supraoptic neurosecretory neurons. By analogy, a similar mode of action may be possible as ANG II facilitates on PRL secretion from the tilapia pituitary. In the tilapia, as in many other teleost fish, PRL cells, comprising nearly 100% of the rostral pars distalis of the pituitary, can be dissected and yield in a nearly homogeneous endocrine tissue. Thus, tilapia PRL cells will provide an ideal model system to analyse the effects of ANG II on responses to osmotic stimulation (Grau and Helms, 1990; Grau *et al.*, 1994). Experiments using dispersed PRL cells of tilapia are now in progress in our laboratory, by simultaneously measuring PRL release and intracellular Ca^{2+} concentrations in perifusion cultures.

Acknowledgements

We are grateful to Prof. Milton H. Stetson, Department of Biological Sciences, University of Delaware, and Dr N. Harold Richman III, Dr Malia M. Chow, Mr Steven K. Shimoda and Ms Claire Ball, Hawaii Institute of Marine Biology, University of Hawaii, for their suggestions and encouragement during the course of this study. Technical assistance of Mr Steve M. Eckert is gratefully acknowledged. We are also indebted to Prof. Yoshio Takei, Ocean Research Institute, University of Tokyo, for his invaluable suggestions and also for his kind provision of ANG II. This study is funded by grants from the University of Hawaii Sea Grant College Program (No. NA86RG0041) and from the United States Department of Agriculture (#983506644).

References

Ayson, F.G., Kaneko, T., Tagawa, M., Hasegawa, S., Grau, E.G., Nishioka, R.S., David S.K., Bern, H.A. and Hirano, T. (1993) Effects of acclimation to hypertonic environment on plasma and pituitary levels of two prolactins and growth hormone in two species of tilapia, *Oreochromis mossambicus* and *Oreochromis niloticus*. *Gen. Comp. Endocrinol.* **89:** 138–148.

Barry, T.P. and Grau, E.G. (1985) Estradiol-17β and thyrotropin-releasing hormone stimulate prolactin release from the pituitary gland of a teleost fish *in vitro*. *Gen. Comp. Endocrinol.* **62:** 306–314.

Bourque, C.W. (1998) Osmoregulation of vasopressin neurons: a synergy of intrinsic and synaptic processes. *Prog. Brain Res.* **119:** 59–76.

Brown, J.A., Paley, R.K., Amer, S. and Aves, S.J. (2000) Evidence for an intrarenal renin–angiotensin system in the rainbow trout, *Oncorhynchus mykiss*. *Am. J. Physiol.* **278:** R1685–R1691.

Brown, P.S. and Brown, S.C. (1987) Osmoregulatory actions of prolactin and other adenohypophysial hormones. In: *Vertebrate Endocrinology: Fundamentals and Biomedical*

Implications (eds P.K.T. Pang, M.P. Schreibman and W.H. Sawyer). Academic Press, London, pp. 45–84.

Canonico, P.L. and MacLeod, R.M. (1986) Angiotensin peptides stimulate phophoinositide breakdown and prolactin release in anterior pituitary cells in culture. *Endcocrinology* **118:** 223–228.

Carrick, S. and Balment, R.J. (1983) The renin–angiotensin system and drinking in the euryhaline flounder, *Platichthys flesus. Gen. Comp. Endocrinol.* **51:** 423–443.

Carroll, R.G., Opdyke, D.F. and Keller, N.E. (1984) Vascular recovery following hemorrhage in the dogfish shark *Squalus acanthias. Am. J. Physiol.* **246:** R825–R258.

Chakfe, Y. and Bourque, C.W. (2000) Excitatory peptides and osmotic pressure modulate mechanosensitive cation channels in concert. *Nature Neuroscience* **3:** 572–579.

Diaz-Torga, G., Iglesias, A.G., Achaval-Zaia, R., Libertun, C. and Becu-Villalobos, D. (1998) Angiotensin II-induced Ca^{2+} mobilization and prolactin release in normal and hyperplastic pituitary cells. *Am. J. Physiol.* **274:** E534–E540.

Eckert, S.M., Yada, T., Shepherd, B.S., Stetson, M.H., Hirano, T. and Grau, E.G. (2001) Hormonal control of osmoregulation in the channel catfish, *Ictarulus punctatus. Gen. Comp. Endocrinol.* **122:** 270–286.

Fitzsimons, J.T. (1998) Angiotensin, thirst and sodium appetite. *Physiol. Rev.* **78:** 583–686.

Fuentes, J. and Eddy, F. (1997) Effect of manipulation of the renin–angiotensin system in control of drinking in juvenile Atlantic salmon (*Salmo salar* L) in fresh water and after transfer to sea water. *J. Comp. Physiol. B* **167:** 438–443.

Grau, E.G. and Helms, M.H. (1990) The tilapia prolactin cell—twenty-five years of investigation. In: *Progress in Comparative Endocrinology* (eds A. Epple, C.G. Scanes and M.H. Stetson). Wiley-Liss, New York, Vol. 342, pp. 534–540.

Grau, E.G., Nishioka, R.S. and Bern, H.A. (1982) Effects of somatostatin and urotensin II on tilapia pituitary prolactin release and interactions between somatostatin, osmotic pressure, Ca^{++}, and adenosine 3′,5′-monophosphate in prolactin release *in vitro. Endocrinology* **110:** 910–914.

Grau, E.G., Nishioka, R.S., Young, G. and Bern, H.A. (1985) Somatostatin-like immunoreactivity in the brain and pituitary of three teleost fish species. *Gen. Comp. Endocrinol.* **59:** 350–357.

Grau, E.G., Richmann, N.H., III and Borski, R.J. (1994) Osmoreception and a simple endocrine reflex of the prolactin cell of the tilapia *Oreochromis mossambicus*. In: *Perspectives in Comparative Endocrinology* (eds K.G. Davey, R.E. Peter and S.S. Tobe), National Research Council of Canada, Ottawa, pp. 251–256.

Hackenthal, E., Paul, M., Ganten, D. and Taugner, T. (1990) Morphology, physiology and molecular biology of renin secretion. *Physiol. Rev.* **70:** 1067–1115.

Helms, L.M.H., Grau, E.G. and Borski, R.J. (1991) Effects of osmotic pressure and somatostatin on the cAMP messenger system of the osmosensitive prolactin cell of a teleost fish, the tilapia (*Oreochromis mossambicus*). *Gen. Comp. Endocrinol.* **83:** 111–117.

Hirano, T. (1986) The spectrum of prolactin action in teleosts. In: *Comparative Endocrinology: Developments and Directions* (ed. C.H. Ralph). A.R. Liss, New York, pp. 53–74.

Hirano, T. and Hasegawa, S. (1984) Effects of angiotensin II and other vasoactive substances on drinking in the eel, *Anguilla japonica. Zool. Sci.* **1:** 106–113.

Iglesias, A.G., Diaz-Torga, G., Piroli, G., Achaval-Zaia, R., De Nicola, A.F., Libertun, C. and Becu-Villalobos, D. (2000) Bromocriptine restores angiotensin II response in pituitary hyperplasia. *Mol. Cell. Endocrinol.* **165:** 67–74.

Kobayashi, H. and Takei, Y. (1996) *The Renin–Angiotensin System. Comparative Aspects.* Springer, Berlin, Heidelberg, New York.

Kobayashi, H., Takei, Y., Itatsu, N., Ozawa, M. and Ichinohe, K. (1983) Drinking induced by angiotensin II in fishes. *Gen. Comp. Endocrinol.* **49:** 295–306.

Malarkey, W.B., Zvara, B.J. and DeGroff, V.L. (1987) Angiotensin II promotes prolactin release from normal human anterior pituitary cell cultures in a calcium-dependent manner. *J. Clin. Endocrinol. Metab.* **64:** 713–717.

Malvin, R.L. (1980) Angiotensin and drinking rate in the euryhaline killifish. *Am. J. Physiol.* **249:** R31–R34.

McCormick, S.D. (2001) Endocrine control of osmoregulation in teleost fish. *Am. Zool.* **41:** 781–791

Naden, R.P., Coultrup, S., Arant, B.S. and Rosenfelt, C.R. (1985) Metabolic clearance of antiotensin II in pregnant and nonpregnant sheep. *Am. J. Physiol.* **249:** E49–E55.

Nishimura, H., Lunde, L.G. and Zucker, A. (1979) Renin response to hemorrhage and hypotension in the aglomerula toadfish *Opsanus tau*. *Am. J. Physiol.* **237:** H105–H111.

Nishioka, R.S., Kelley, K.M. and Bern, H.A. (1988) Control of prolactin and growth hormone secretion in teleost fishes. *Zool. Sci.* **5:** 267–280.

Ogilvy, C.S., Tremml, P.G. and DuBois, A.B. (1988) Pressor and hemodilution responses compensate for acute hemorrhage in bluefish. *Comp. Biochem. Physiol. A* **91:** 807–813.

Okimoto, D.K., DiStefano, J.J., Kuwaye, T.T., Ron, B., Weber, G.M., Nguyen, T.T. and Grau, E.G. (1994) On plasma volume measurement and the effect of experimental stress in the male tilapia, *Oreochromis mossambicus*, maintained in fresh water. *Fish Physiol. Biochem.* **12:** 431–438.

Rentier-Delrue, F., Swennen, D., Prunet, P., Lion, M. and Martial, J.A. (1989) Tilapia prolactin: Molecular cloning of two cDNAs and expression in *Escherichia coli*. *DNA* **3:** 261–270.

Russell, M.J., Klemmer, A.M. and Olson, K.R. (2001) Angiotensin signaling and receptor types in teleost fish. *Comp. Biochem. Physiol. A* **128:** 41–51.

Saavedra, J.M. (1992) Brain and pituitary angiotensin. *Endo. Rev.* 13: 329–380.

Seale, A.P., Itoh, T., Moriyama, S., Takahashi, A., Kawauchi, H., Sakamoto, T., Fujimoto, M., Riley, L.G., Hirano, T. and Grau, E.G. (2002) Isolation and characterization of a homologue of mammalian prolactin-releasing peptide from the tilapia brain and its effect on prolactin release from the tilapia pituitary. *Gen. Comp. Endocrinol.* **125:** 328–339.

Shepherd, B.S., Sakamoto, T., Hyodo, S., Ball, C., Nishioka, R.S., Bern, H.A. and Grau, E.G. (1999) Is the basic regulation of pituitary prolactin ($tPRL_{177}$ and $tPRL_{188}$) secretion and gene expression in the euryhaline tilapia (*Oreochromis mossambicus*) hypothalamic or environmental? *J. Endocrinol.* **161:** 121–129.

Specker, J.L., King, D.S., Nishioka, R.S., Shirahata, K., Yamaguchi, K. and Bern, H.A. (1985) Isolation and partial characterization of a pair of prolactins released *in vitro* by the pituitary of the cichlid fish, *Oreochromis mossambicus*. *Proc. Natl Acad. Sci. USA* **82:** 7490–7494.

Takei, Y. (1988) Changes in blood volume after alteration of hydromineral balance in conscious eels, *Anguilla japonica*. *Comp. Biochem. Physiol.* **91A:** 293–297.

Takei, Y., Kobayashi, H. and Hirano, T. (1979) Angiotensin and water intake in the Japanese eel, *Anguilla japonica*. *Gen. Comp. Endocrinol.* **38:** 466–475.

Tierney, M.L., Luke, G., Cramb, G. and Hazon, N. (1995) The role of the rennin–angiotensin system in the control of blood pressure and drinking in the European eel, *Anguilla anguilla*. *Gen. Comp. Endocrinol.* **105:** 9–17.

Tsuchida, T. and Takei, Y. (1998) Effects of homologous atrial natriuretic peptide on drinking and plasma ANG II level in eels. *Am. J. Physiol.* **275:** R1605–R1610.

Tsuchida, T. and Takei, Y. (1999) A potent dipsogenic action of homologous angiotensin II infused at physiological doses in eels. *Zool. Sci.* **16:** 479–483.

Weber, G.M., Powell, J.F.F., Park, M., Fischer, W.H., Craig, A.G., Rivier, J.E., Nanakorn, U., Parhar, I.S., Ngamvongchon, S., Grau, E.G. and Sherwood, N.M. (1997) Evidence that gonadotropin-releasing hormone (GnRH) functions as a prolactin-releasing factor in a teleost fish (*Oreochromis mossambicus*) and primary structures for three native GnRH molecules. *J. Endocrinol.* **155:** 121–132.

Wigham, T., Nishioka, R.S. and Bern, H.A. (1977) Factors affecting *in vitro* activity of prolactin cells in the euryhaline teleost *Sarotherodon mossambicus (Tilapia mossambica). Gen. Comp. Endocrinol.* **32:** 120–131.

Yada, T., Hirano, T. and Grau, E.G. (1994) Changes in plasma levels of the two prolactins and growth hormone during adaptation to different salinities in the euryhaline tilapia, *Oreochromis mossambicus. Gen. Comp. Endocrinol.* **93:** 214–223.

Yamaguchi, K., Specker, J.L., King, D.S., Yokoo, Y., Nishioka, R.S., Hirano, T. and Bern, H.A. (1988) Complete amino acid sequences of a pair of fish (tilapia) prolactins, $tPRL_{177}$ and $tPRL_{188}$. *J. Biol. Chem.* **263:** 9113–9121.

8

Nephron structure and immunohistochemical localization of ion pumps and aquaporins in the kidney of frogs inhabiting different environments

M. Uchiyama and H. Yoshizawa

1. Introduction

At the end of the Devonian period of the Paleozoic era, 370 million years ago, the first amphibians appeared on the earth. The early amphibians needed to adapt to the new terrestrial environment with regard to regulation of the water and electrolyte contents of their bodies, the excretion of metabolic waste and the exchange of respiratory gases. Many of the modern amphibians (the name derives from the Greek *amphi* 'both kind' and *bios* 'life') spend part of their lives in water and part on land, and they inhabit niches ranging from aqueous (fresh water (FW) and sea-water (SW)) to arid terrestrial (arboreal and desert). In amphibians, rates of water evaporation via the skin are very high in arid conditions; osmotic water fluxes are relatively high in water. Thus, amphibians continuously face the threat of desiccation as a result of water loss across their highly permeable skin on land and in air; on the other hand, amphibians experience the opposite problems of osmotic flooding and diffusive and urinary losses of electrolytes. Therefore, transfer between the two environments results in a diminution and concentration of body fluids on land, and expansion and dilution of body fluids in water. How do they solve these problems physiologically, which integrated mechanisms were developed to cope with these problems? It is well-known that in amphibians the skin, the kidney and the urinary bladder are important osmoregulatory organs whose functions are controlled by both neural and endocrine systems.

Osmoregulation and Drinking in Vertebrates, edited by N. Hazon & G. Flik.
© 2002 BIOS Scientific Publishers Ltd, Oxford

We review in short the structures of these three organs in relation to habitats of anuran amphibians. Amphibians do not drink orally but instead absorb water and ions through the ventral skin from water and wet substrate. Many studies showed control mechanisms for exchanges of ions and water across the skin (Fox, 1987; Parsons, 1994). In the amphibian skin, two main types of cells are distinguished. The predominant granular cells, along with the underlying spiny and germinal cells, constitute a syncytium of principal cells (PC), which are the main sites of Na^+ transport. Ouabain-binding sites, as a measure of the Na^+,K^+-ATPase, the enzyme that mediates active extrusion of Na^+ from the intra-cellular pool, were identified in all these cells (Mills *et al.*, 1977). The other, much less abundant and mitochondria-rich cells (MRC) have apical Cl^- channels, which provide a trans-cellular diffusion pathway for chloride ions (Larsen, 1991). H^+-ATPase (vacuolar-ATPase) was localized to the apical membrane of MRC (Ehrenfeld and Klein, 1997). It is supposed that active transport of Cl^- is energized by a proton pump, and the amphibian skin can transport sodium and chloride into the body fluid that is 200–500 times more concentrated in these ions than FW when the frog is submerged (Jensen *et al.*, 1997). Furthermore, frog skin cells possess Cl^-/HCO_3^- antiporter and amiloride-sensitive epithelial sodium channels (ENaC; Larsen, 1991; Puoti *et al.*, 1997). Usually terrestrial species have higher osmotic permeability than do aquatic species, which has been interpreted as a secondary adaptation to aquatic environments (see review Boutilier *et al.*, 1992). For most amphibians on land, 'cutaneous drinking' through the ventral skin is extremely important since rates of evaporative water loss via the skin are very high and these animals do not drink by the mouth (e.g. Bentley and Yorio, 1979). Movement of water across the skin is a passive process; the coupling of transport of water and ions across frog skin may be secondary to a current-induced local osmotic gradient and the osmotic water permeability of the epithelium. It seems that there are two pathways for water movement across the epithelium, one is the para-cellular pathway and the other is a trans-cellular pathway, which requires the operation of water channels (aquaporins, AQPs). Water movements across the cell membranes are postulated to occur via the 'major intrinsic protein' (MIP) that forms channels for water movements. In the skin and the urinary bladder of toads and frogs, a water channel was demonstrated by molecular cloning with close homology to mammalian AQP-1 (Abrami *et al.*, 1994; Siner et al., 1996). Recently, two distinct cDNAs encoding water channel proteins were identified from the ventral skin of Japanese tree frogs (Tanii *et al.*, 2000). Alignments of these cDNAs show more homology to mammalian AQP-3. Christensen (1974) reported on the important role of the vasculature and the perfusion of the blood vessels of the abdominal skin of the toad *Bufo bufo* for water fluxes. Thus studying amphibian skin physiology with respect to water and ion transport should also consider the regulation of the vascular bed. The high rate of water flux across the isolated skin may depend on a high osmotic gradient across the skin. Control mechanisms of water uptake are well-studied and the neurohypophysial hormone arginine vasotocin (AVT), epinephrine and angiotensin II (ANG II) are likely candidates to stimulate the water absorption response. The osmotic permeability of the skin is correlated with habitat selection (see review, Bentley, 1971; Mullen and Alvarado, 1976), and the osmotic gradient across the skin determines the rate of water entry. Higher permeability is generally found in skin of terrestrial species, lower permeability in more aquatic species.

The amphibian urinary bladder, like the skin, is a widely used *in vitro* preparation for studies of sodium transport mechanisms and control of epithelial water permeability (see Bentley, 1966; Bentley, 1971). The organ stores not only water but also minerals and urea, and AVT stimulates active sodium uptake and water conservation from the bladder (Bentley, 1966; Bentley, 1971). In an early ultrastructural study of the toad urinary bladder, water channels contained within 'intramembrane particle aggregates' were the bases for the shuttle hypothesis suggesting that antidiuretic hormone (ADH)-sensitive water permeability was regulated by trafficking of specialized vesicles, called aggrephores (Muller and Kachadorian, 1984). The ADH-sensitive aggrephore may represent an AQP shuttle mechanism. Although so far, no AQP homologous to mammalian AQP-2 has been reported in sub-mammalian species including amphibians, toads and frogs do reabsorb water from their urinary bladders when they are dehydrated (Shoemaker, 1964). Bentley (1966) reported that terrestrial and arboreal species (e.g. *Bufo* and *Hyla*) hold 25–50% of their standard mass as dilute urine in the bladder, while aquatic anurans (e.g. *Xenopus*) have very small bladder capacities (1–5%). Thus, the two important osmoregulatory organs, the skin and the urinary bladder, are structurally significant correlates to the habitat in amphibians.

Although there are two excellent review books on comparative physiology of the vertebrate kidney (Dantzler, 1989; Braun and Dantzler, 1997), renal structures and functions have not been fully defined in amphibians adapted to various environments. Amphibian kidneys are unable to produce urine hyper-osmotic to the plasma, but they do produce urine varying from iso-osmotic with the plasma during dehydration to one-tenth the osmolality of the plasma during hydration (see review Braun and Dantzler, 1997). It has been known that the renal excretory capacities of aquatic and terrestrial amphibians did not differ markedly when they were kept in water (Shoemaker and Nagy, 1977). They produce dilute urine and conserve electrolytes while excreting excess fluid in water and rapidly reduce or cease urine production when again out of the water. Thus, water conservation prevails over renal excretion. The volume and the composition of urine are determined by the glomerular filtration rate (GFR), and reabsorption and secretion of water and solutes in the nephron. In amphibians out of water, reduction or cessation of urine production is essential to conserve water. This response seems to occur even without significant dehydration or elevation of plasma osmolality in all amphibians examined. This antidiuretic response is mediated by hormonal and neuronal controls (see reviews e.g. Dantzler, 1989; Shoemaker *et al.*, 1992). Many studies have been done to characterize the location, the molecular nature, and physiological and pharmacological properties of ion pumps, co-transporters and channels in mammalian kidneys. Studies using immunocytochemical methods have been carried out and helped to recognize the functions of the cells that constitute the distinct segments of the nephron. In the nephron segments, the primary functions of reabsorption and secretion are defined by a differential and polarized insertion of specific transporters (and related proteins) into apical membranes or basal domains: for example, apical membranes of the proximal tubule are characterized by the presence of many sodium-coupled transporters and by ENaC. On the other hand, basolateral membranes are characterized by the electrogenic sodium pump, the Na^+,K^+-ATPase.

According to histological studies, the kidney of amphibians is composed of a number of nephrons containing a variety of cell types (Uchiyama *et al.*, 1990;

Mobjerg, *et al.*, 1998). It is assumed that morphological segmentation of the nephron is accomplished by differential expression of membrane proteins, transporters, and channels in amphibians and mammals alike. There are, however, only a few studies on the localization of the ion pumps in the kidney of amphibians studied from a comparative morphological point of view and the evolution of osmoregulation in vertebrates. Therefore, the purposes of the present study are to examine (1) whether the ecological factors influence the structures of the kidneys in anuran amphibians, (2) where the main sites and cells are to transport the major cations. The study was conducted by means of morphological and immunohistochemical methods.

2. Materials and Methods

2.1 *Animals*

Fifteen species of anuran amphibians that inhabit different environments, such as terrestrial, amphibious and aquatic, were used in this study. They are listed in *Table 1*.

2.2 *Histological observations and reconstruction of nephrons*

Toads and frogs were anaesthetized in a 0.3% solution of tricaine methanesulfonate (MS222; Sankyo, Japan). The animals were perfused transcardially with saline followed by a mixture of 4% paraformaldehyde and 0.25% glutaraldehyde in 0.1 M phosphate buffer (pH 7.4) or by Bouin's fixative without acetic acid. In small species, the kidneys were dissected out and were immediately fixed with the fixatives. Abdominal skin was dissected and immediately immersed in fixative. The samples were embedded in epoxy resin or paraffin and cut into 6–8-μm serial sections. Reconstruction of single nephrons was performed from serial cross- and longitudinal sections stained with haematoxylin-eosin. For transmission electron microscopy, tissue was post-fixed with 1% osmium tetroxide for 1 h. One-micrometre sections of Araldite-embedded slices were stained with toluidine blue for light microscopy; ultrathin sections for the transmission electron microscopy were stained with uranyl acetate and alkaline lead citrate, and examined with a JEOL JEM-100B electron microscope. The structural organization and the measurement of nephron segments were based on 2–5 specimens of each species. Under light microscopy the image data of each slice of serial section was captured by means of a digital photo system (Coolpix 950, Nikon, Japan). Using the software for 3D reconstruction of serial-section data (Voxblast light, VayTek Inc., USA), the graphic figure of a single nephron was reconstructed with a PC compatible computer (Endeavor Pro-700L, Epson, Japan).

2.3 *Immunohistochemistry*

The sections were subsequently processed immunohistochemically according to the peroxidase–antiperoxidase (PAP) or the avidin–biotin–horseradish peroxidase complex (ABC) procedures. Sections were counterstained with haematoxylin. Validation of the immunohistochemical procedures was accomplished by using secondary antibody only (primary antiserum was replaced by buffer) or normal rabbit serum to replace primary antiserum.

Table 1. *Anuran amphibians and characterizations of their kidney and nephrons*

Species	Habitat	Body weight (g)	Number of glomeruli	Length of proximal & distal tubules (μm)	Cell size of proximal & distal tubules (μm)	No. of animals
Bufo alvarius	T	215.0	–	– , –	– , –	3
Bufo j. formosus	T	149.1	7869 ± 1341	3262; 3052	14.7; 11.6–12.2	20
Bufo punctatus	T	6.6	–	– , –	– , –	4
Bufo raddei	T	–	–	– , –	– , –	3
Hyla japonica	Arb	1.7	241 ± 53	774; 534	15.6; 9.2–12.2	45
Microhyla ornate	T	3.1	–	– , –	– , –	5
Rana catesbeiana	Am	499.7	–	– , –	– , –	36
Rana japonica	Am	3.9	195 ± 25	– , –	– , –	10
Rana niguromaulata	Am	7.2	477 ± 69	1428; 712	12.7; 7.5–10.0	14
Rana ornativentris	Am	4.7	–	– , –	– , –	2
Rana rugosa	Am	4.1	–	– , –	– , –	4
Bombina orientalis	Am	7.6	–	– , –	– , –	3
Limnonectes cancriovorus	Am (sea water)	41.7	1900	– , –	20.6; 10.0–17.5	4
Xenopus laevis	Aq	66.3	3830 ± 2042	3057; 1109	18.8; 11.4–14.4	5
Hymenochirus boettgeri	Aq	1.3	64 ± 5	– , –	– , –	4

Abbreviations: Am, amphibious; Aq, aquatic; Arb, arboreal; T, terrestrial.

Antisera. Polyclonal anti-Na+,K+-ATPase α-subunit antiserum and polyclonal anti-H^+-ATPase antiserum raised in white rabbits were kindly provided by Profs Yamauchi (Hokkaido Univ., Hakodate, Japan) and Moriyama (Okayama Univ., Okayama, Japan), respectively. The procedures of the antigenic peptide synthesis and immunization were described elsewhere (Ura et al., 1996; Hayashi et al., 2000). Polyclonal anti-H^+,K^+-ATPase-α subunit antibody, rabbit anti-AQP 2 and rabbit anti-AQP 3 affinity-purified polyclonal antisera were purchased (Chemicon International Inc. CA, USA).

3. Morphology of anuran kidney

3.1 *Gross anatomy*

The kidneys of all anurans were similar at the gross morphological level. The paired kidneys are elongated, reddish-brown structures lying beside the dorsal aorta. In toads, the medial margin was indented by grooves running over the dorsal surface to produce six to eight lobes. On the contrary, the ventral and dorsal surfaces of the kidney were smooth in frogs. From correlation analysis, there is a significant relation in the scaling of kidney weight with body mass among animals (r^2 = 0.91; *Figure 1*). The aquatic species, *Xenopus* sp., has long and narrow kidneys in comparison with those of terrestrial or semi-aquatic species. Thus, it seems that the external shape of the kidneys of anuran amphibians varies only slightly among species from different habitats.

Ogawa (1961) suggested that ecological factors might influence the shape of kidneys within fish more than in any other vertebrates group. FW teleosts, which maintain the osmolality of their body fluids above that of their environments, have nephrons that differ from those of marine teleosts. The glomeruli tend to be larger

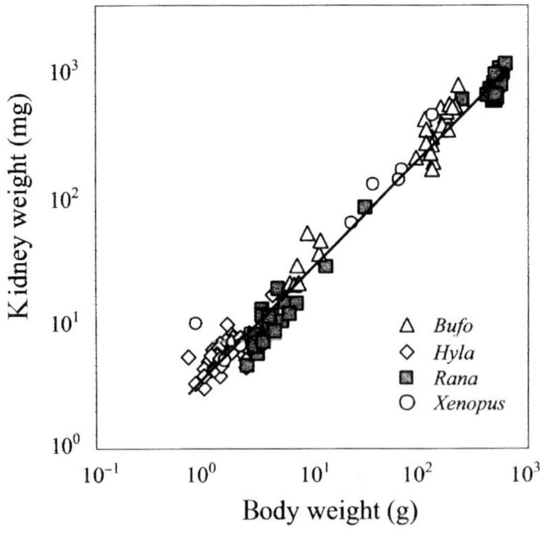

Figure 1. *Allometric relationship and the influence of habitat. Kidney mass in anuran amphibians increases with body mass.* Bufo *sp. and* Hyla *sp. are terrestrial and arboreal species, respectively.* Rana *sp. and* Xenopus *sp. are semi-aquatic and aquatic species, respectively.*

and the nephrons more numerous than in marine teleosts. On the other hand, marine teleosts maintain the osmolality of their body fluids below that of their environments. The kidneys are divided into an anterior head kidney (mainly haematopoietic and endocrine tissues) and posterior trunk kidney containing the true renal tissue, but the two kidney compartments are fused partially or completely. A distal tubule is absent in most species (Hickman and Trump, 1969). In birds and mammals, the urine concentrating mechanism depends on the arrangement of the loop of Henle of the nephron, which functions as a countercurrent multiplier system. Bankir and Rouffignac (1985) underlined the importance of the inner stripe architecture of mammalian renal medulla in urine concentration. Beuchat (1996) concluded that the relationship between urine osmolality and absolute length of the loop of Henle was neither proportional nor direct using data for over 330 species of mammals. Urine concentrating ability was independent of the thickness of the outer medulla in mammals of mesic, arid and FW environments. In marine mammals, the kidneys are large and composed of multiple discrete renicules that accommodate the lengthy tubules (Beuchat, 1996). On the other hand, the kidney of crab-eating frog, *Limnonectes cancrivorus*, that survives in SW, does not show any specialized structures compared with the structure of other frog kidneys (Uchiyama *et al.*, 1990). In the present study, it seems that lobular structure of the kidney of terrestrial toads does not increase parenchymal mass but may have an advantage to increase the length of tubules.

3.2 *Morphology*

According to observations by light and electron microscopy, amphibian kidneys are divided into central (the ventromedial) and peripheral (the dorsolateral part of the kidney) zones. The former contains only distal tubules, while the latter contains the other part of nephrons; the glomeruli are situated in a distinct band between the two zones. In all frogs and toads examined, the nephrons have eight identifiable components, viz. a Bowman's capsule in the renal corpuscle, a neck segment, a proximal tubule, an intermediate segment, an early distal tubule, a late distal tubule, a connecting tubule, and a collecting duct. The proximal tubule cells of anuran amphibians have brush borders and numerous mitochondria, but lack deep interdigitations of basolateral membranes with mitochondria (*Figure 2A1, 2B1*). Numerous peroxysomes are found. The cells of early distal tubules at the end of the thin intermediate segment have deep basolateral infoldings with elongated mitochondria occupying the cytoplasm between these infoldings. The cells of late distal tubules were taller and have much deeper basal infoldings and much larger nuclei than the cells of early distal tubules (*Figure 2A2, 2C1*). The connecting tubule in the terrestrial and amphibious species consists primarily of light principal cells with some dark mitochondria-rich intercalated cells. From the transition of the distal tubule to the collecting duct toward the end of the distal tubule the number of intercalated cells is gradually increasing (*Figure 2B2*). In connecting tubules and collecting ducts of the aquatic species, *X. laevis* and *Hymenochirus boettgeri*, canaliculus cells (also called 'flask cells') were observed (*Figure 2C2*). The neck and the intermediate segments consist of common cells characterized by a low columnar shape and a multiciliated apex. These observations confirm previous observations on the kidney of amphibians (Uchiyama *et al.*, 1990; Mobjerg *et al.*, 1998). Numbers of nephrons per kidney

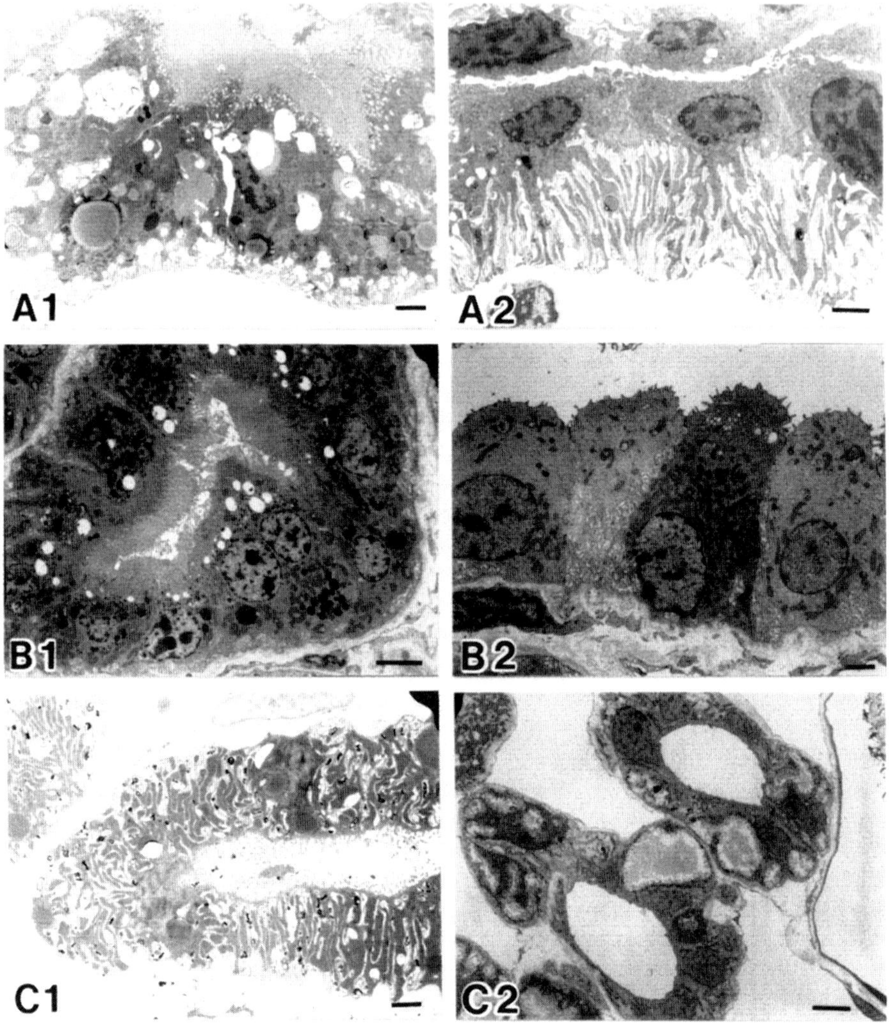

Figure 2. *Electron micrographs of the kidney cells in three anuran species,* Hyla japonica *(A),* Bufo j. formosus *(B) and* Xenopus laevis *(C). A1, Proximal tubule cells. A2, Early distal tubule cells. Scale bars = 2 µm each. B1, Proximal tubule, scale bar = 2 µm. B2, Connecting tubule cells, the intercalated dark cell with abundant mitochondria and the principal cells with a pale cytoplasm. Scale bar = 5 µm. C1, Early distal tubule. Scale bar = 2 µm. C2, Connecting tubule cells with vacuolated and/or canaliculi cytoplasm. Scale bar = 7 µm.*

were estimated at about 8000, 241, 500, 1900, 60 and 3800 in *B. j. formosus, H. japonica, R. nigromaculata, L. cancrivorus, Hym. boettgeri* and *X. laevis*, respectively. Average diameters of Bowman's space were 98, 59, 59, 65, 92 and 110 µm in *B. j. formosus, H. japonica, R. japonica, R. nigromaculata, Hymenochirus boettgeri* and *X. laevis*, respectively. The diameters of Bowman's space of terrestrial frogs were smaller than those of aquatic species. The diameters were 118 and 123 µm in *L. cancrivorus* acclimated to 50% SW and FW, respectively. Lengths of the proximal and distal (early distal to connecting) tubules were 3.2 and 3.0 mm, respectively, in *B. j.*

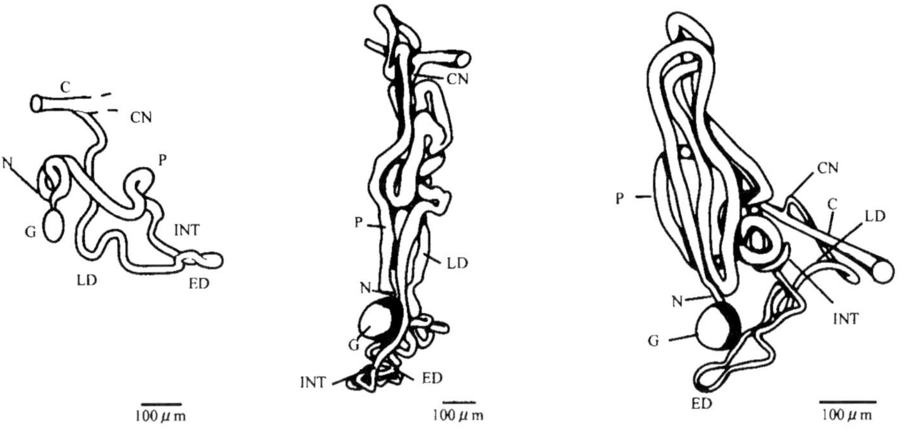

Figure 3. *Schematic reconstruction of nephrons indicating the position of the various segments in three anuran species,* Hyla japonica *(A),* Bufo j. formosus *(B) and* Xenopus laevis *(C). Abbreviations: C, collecting duct; CN, connecting tubule; ED, early distal tubule; G, glomerulus; INT, intermediate segment; N, neck segment; LD, late distal tubule; P, proximal tubule.*

formosus (*Figure 3B*). On the other hand, those lengths were 3.0 and 1.1 mm respectively in *X. laevis* (*Figure 3C*). These data are shown in *Table* 1. Although the kidneys of all anurans were similar at gross morphological level, the structure of the nephrons was related to the habitat. In the aquatic species, a large volume of plasma is filtrated in some large glomeruli and the ultrafiltrate is reabsorbed in a large and long proximal segment of the nephron. Control of tubular transport may be less-developed when a nephron is characterized by a small and short distal segment. Terrestrial species have nephrons with a long well-developed distal segment and regulatory mechanisms of tubular transport may exist in these segments. Thus, it seems that development of the late distal segments of the nephron is an important factor for adaptation to a terrestrial habitat in anurans.

4. Na^+,K^+-ATPase

4.1 *Immunohistochemical studies of Na^+,K^+-ATPase*

The anuran amphibians used in this study and the cross-reactivity of the antisera used for detection of ion pumps are shown in *Table 2*. Immunohistochemical observations and diagrammatic representations are shown in *Figures 4A and 5A*. The basic structure of the nephron and the distribution pattern of Na^+,K^+-ATPase activities were identical in the kidneys of all species examined. Immunoreactivity was obtained with anti-Na^+,K^+-ATPase antisera in various parts of the nephron. In the proximal tubule two parts are recognized, a straight and a convolute. In the proximal convoluted tubule, the basal part of the basement membrane of tall columnar cells with a dense brush border was immunoreactive. Apical and lateral parts of the cells were not stained in this section. The immunoreactivity for Na^+,K^+-ATPase was lower in proximal straight tubules than in proximal convoluted tubules. In the distal

Table 2A. *Localization of Na^+,K^+-ATPase along the nephron of the kidney in anuran amphibians*

	Neck segment	Proximal tubule	Intermediate segment	Early distal tubule	Late distal tubule	Connecting tubule	Collecting duct
Bufo alvarius	–	+	–	++	++	+	–
Bufo raddei	–	+	–	++	++	+	–
Bufo j. formosus	–	+	–	++	++	+	+
Hyla japonica	–	+	–	++	++	+	+
Rana japonica	–	+	–	++	++	+	+
Rana ornativentris	–	+	–	++	++	+	+
Rana niguromaulata	–	+	–	++	++	+	+
Hymenochirus boettgeri	–	+	–	+	+	+	+
Xenopus laevis	–	+	–	+	+	+	+

Table 2B. *Localizations of H^+-ATPase along the nephron of the kidney in anuran amphibians*

	Neck segment	Proximal tubule	Intermediate segment	Early distal tubule	Late distal tubule	Connecting tubule	Collecting duct
Bufo alvarius	–	–	–	+	+	+	+
Bufo raddei	–	–	–	+	+	+	+
Bufo j. formosus	–	–	–	–	–	+	+
Hyla japonica	–	–	–	–	–	+	+
Rana japonica	–	–	–	–	–	+	+
Rana ornativentris	–	–	–	–	–	+	+
Rana niguromaulata	–	–	–	–	–	+	+
Hymenochirus boettgeri	–	+	–	+	+	+	+
Xenopus laevis	–	+	–	+	+	+	+

Table 2C. *Localization of H^+,K^+-ATPase along the nephron of the kidney in anuran amphibians*

	Neck segment	Proximal 1 tubule	Intermediate segment	Early distal tubule	Late distal tubule	Connecting tubule	Collecting duct
Bufo alvarius	–	–	–	–	–	+	+
Bufo raddei	–	–	–	+	+	+	+
Bufo j. formosus	–	–	–	+	+	+	+
Hyla japonica	–	–	–	+	+	+	+
Rana japonica	–	–	–	+	+	+	+
Rana ornativentris	–	–	–	+	+	+	+
Rana niguromaulata	–	–	–	+	+	+	+
Hymenochirus boettgeri	–	–	–	+	+	+	+
Xenopus laevis	–	+	–	+	+	+	+

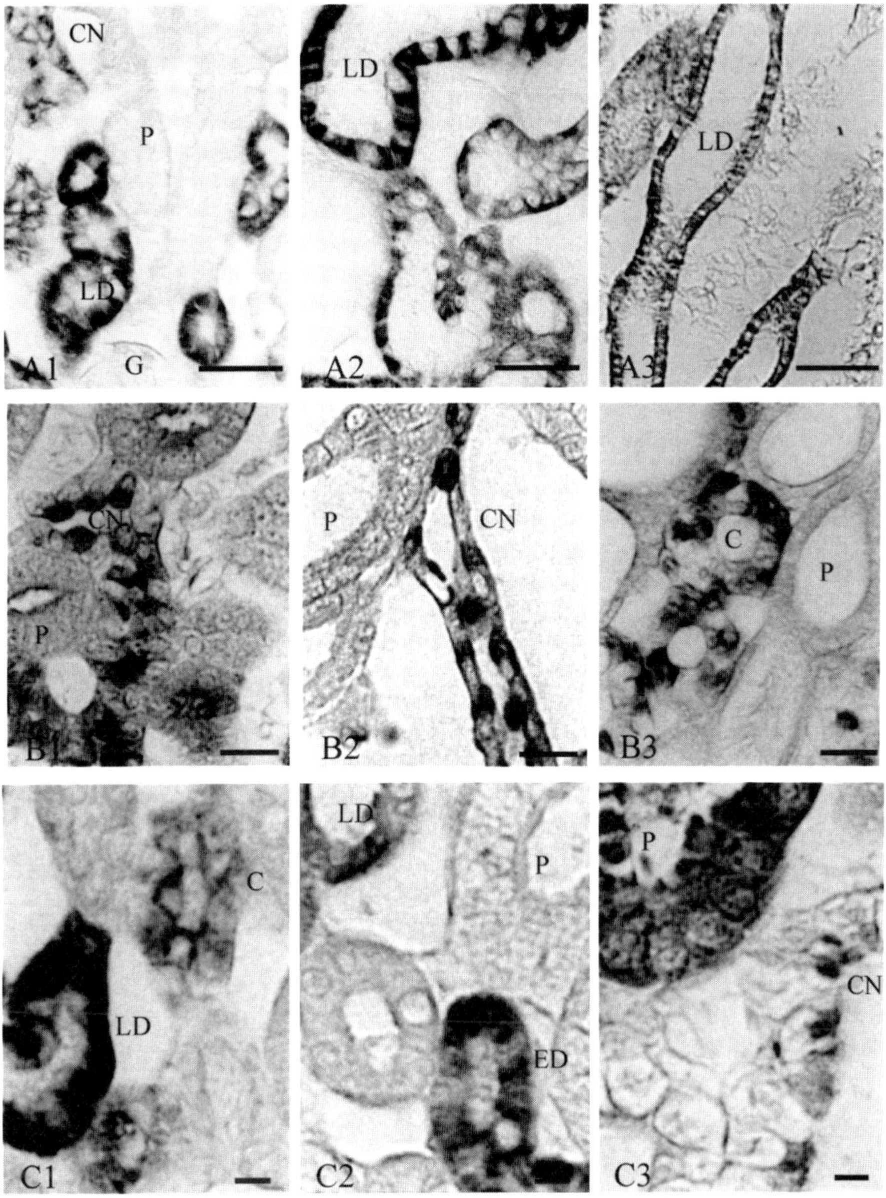

Figure 4. *Light micrographs of paraffin sections from anuran amphibian kidneys processed for immunocytochemistry to reveal* Na^+,K^+*-ATPase,* H^+*-ATPase and* H^+*-*K^+*-ATPase immunoreactivities employing polyclonal antibodies. (A) Immunohistochemical localizations of* Na^+,K^+*-ATPase in late distal tubules of* Hyla japonica *(A1),* Bufo j. formosus *(A2) and* Xenopus laevis *(A3). Scale bar=50 μm each. (B) Immunohistochemical localizations of* H^+*-ATPase in connecting tubules of* Hyla japonica *(B1),* Bufo j. formosus *(B2) and* Xenopus laevis *(B3). Scale bar=20 μm each. Immunohistochemical localizations of* H^+,K^+*-ATPase in early distal, late distal tubules and connecting tubules of* Hyla japonica *(C1),* Bufo j. formosus *(C2) and* Xenopus laevis *(C3). Scale bar=10μm each. Abbreviations: see legend to Figure 3.*

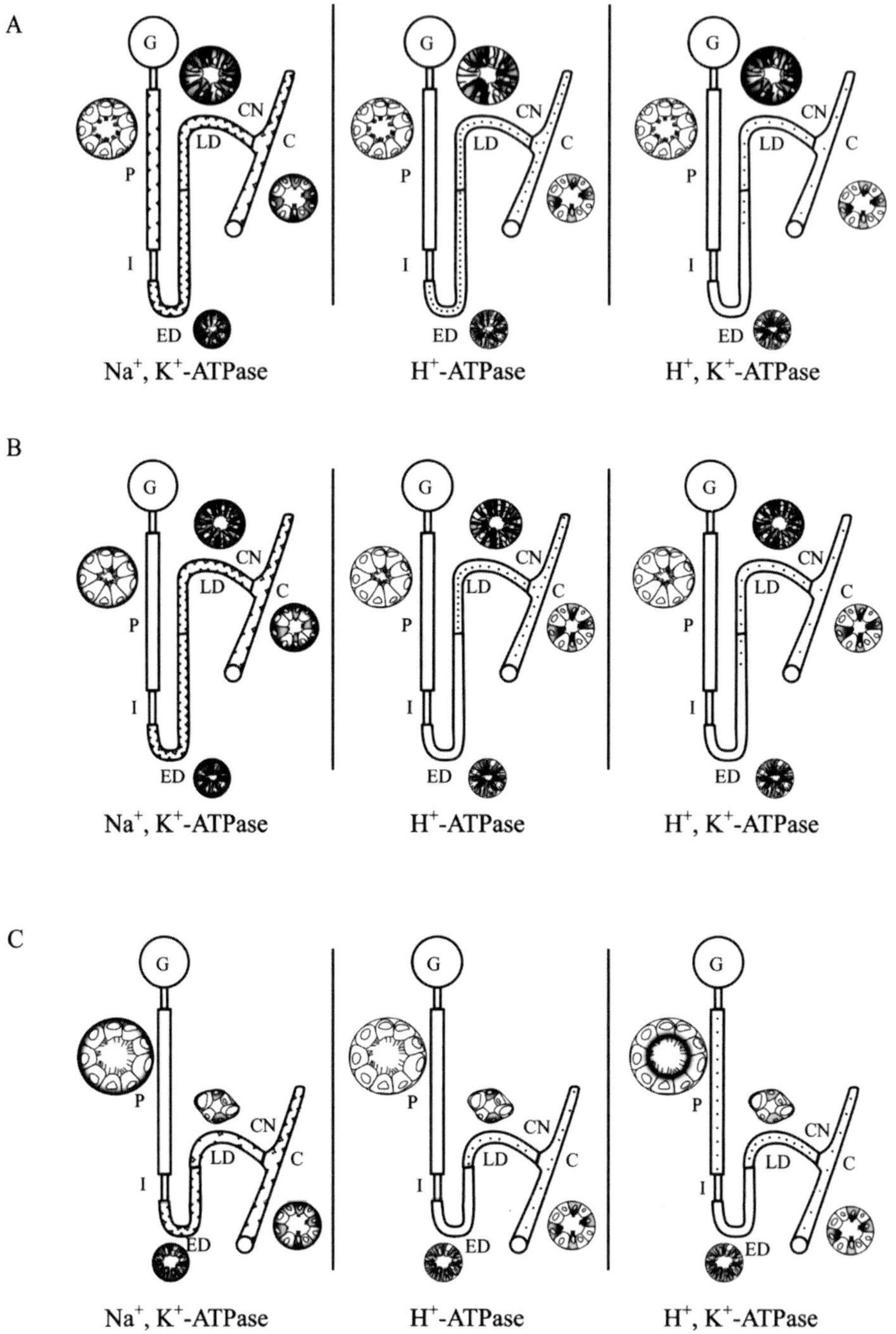

Figure 5. *Diagrammatic representations of nephrons and localization of* Na^+,K^+*-ATPase,* H^+*-ATPase and* H^+,K^+*-ATPase, along the nephron of anuran amphibians. (A) Schematic nephrons of* Hyla japonica. *(B) Schematic nephrons of* Bufo j. formosus. *(C) Schematic nephrons of* Xenpus laevis. *Abbreviations: see legend to Figure 3.*

tubules, the early distal tubule consists of heavily interdigitated cells and the elongated interdigitating cytoplasmic processes reach close to the luminal border, and this interdigitating cytoplasm was well-immunoreactive. Immunostaining was found basally and laterally of the nucleus in all terrestrial and semi-terrestrial species, but not in aqueous species. In the connecting tubules and collecting ducts, basolateral membranes were immunoreactive in all species. In the renal corpuscle, the neck segments and the intermediate segments, no immunoreactivity was observed in all species examined. When Na^+,K^+-ATPase activity is measured biochemically in discrete nephron segments dissected from mammalian kidney, the highest activities are measured in the thick ascending limb and distal convoluted tubule (Doucet *et al.*, 1979; Hersey *et al.*, 1985). In the thick ascending limb segment high rates of transport of Na^+, K^+ and Cl^- ions occur, but these segments are water-impermeable in mammals. Thus these segments are not only involved in the formation of dilute urine, but also enable the mammalian kidney to concentrate urine by supplying Na^+ and Cl^- ions to the renal medulla, thus providing energy for countercurrent multiplication. The non-mammalian convoluted distal tubule or diluting segment has a remarkable resemblance to that of the mammalian thick ascending limb. The cell in this segment has a large nucleus and abundant mitochondria and the basal cell membrane is highly interdigitated (Guggino *et al.*, 1988). The immunohistochemistry presented here shows that the highest Na^+,K^+-ATPase activity resides in the early distal tubule. In amphibians, this segment conserves Na^+ and Cl^- while it simultaneously produces hypotonic urine. The major sites of Na^+ and Cl^- transport in the renal tubules, determined by micropuncture or microperfusion techniques, vary among the vertebrates. The proximal tubule has generally been considered the primary site for the absorption of filtered Na^+ and Cl^- in mammals and birds. In amphibians, however, only some 20–45% of the filtered sodium is absorbed along the proximal tubules and some 50–70% of the filtered sodium is absorbed along the distal tubules or collecting ducts (Long, 1973; Stoner, 1985). In amphibians, therefore, the occurrence of Na^+,K^+-ATPase along the nephrons is comparable to that in mammalian nephrons and thus may play a similar important role. The present study would further suggest that there is only minor sodium absorption in the neck and intermediate segments because of a virtual absence of Na^+,K^+-ATPase.

4.2 *Hormonal regulation of Na^+,K^+-ATPase activity*

Na^+,K^+-ATPase is subject to both short- and long-term regulation by a variety of hormones. They are: steroid hormones, catecholamines and peptide hormones. The corticosteroid aldosterone plays an important role in Na^+ and K^+ transport in the kidney (Feraill and Doucet, 2001). The main effect of aldosterone on Na^+,K^+-ATPase is to sustain long-term increased expression of the sodium pump (Welling *et al.*, 1993). Chronic effects of aldosterone include an increase in the turnover rate of the basolateral Na^+,K^+-ATPase, insertion of new Na^+,K^+-ATPase units, and increased permeability to both sodium and potassium of the apical membrane (Verrey *et al.*, 1996). Catecholamines affect Na^+,K^+-ATPase activity, and noradrenalin and dopamine are the two best-studied regulators. They act antagonistically in regulating salt reabsorption in the mammalian kidney (see reviews, Aperia *et al.*, 1996; Meister and Aperia, 1993). In the kidney, dopamine inhibits the Na^+,K^+-ATPase activity and this represents an important mechanism for regulating salt reabsorption during a high salt intake (Baines *et al.*, 1992). Adrenalin and noradrenalin stimulate sodium pump activity and appear to have a role in sodium absorption in the nephron (see review Aperia *et al.*, 1996). Peptide hormones form another group of essential Na^+,K^+-ATPase regulators. Insulin stimulates the enzyme activity and increases potassium uptake (see Sweeney and Klip, 1998). In the collecting tubule of rat kidney, insulin increased the affinity of the enzyme for sodium ions. In mammalian kidney, other hormones that modulate the enzyme activity are angiotensin II (ANG II) (Buhagiar *et al*, 1999), vasopressin (Wilson and Horster, 1983) and atrial natriuretic peptides (ANPs) (Beltowski *et al.*, 1998), which are known as hydromineral hormones. As yet little is known about effects of these regulators on Na^+,K^+-ATPase activity in amphibian kidney.

5. H^+-ATPase

5.1 *Immunohistochemical localization of H^+-ATPase*

Anuran amphibians used in this study and the immunoreactivities against antibodies of the ion pumps are shown in *Table* 2. The distribution pattern of H^+-ATPase was almost identical in the kidney of all species examined in the present study (*Figures* 4B, 5). In the proximal tubules, apical membranes of the cells were slightly immunoreactive in aquatic species, *X. laevis* and *Hymenochirus boettgeri*. In the other species examined, no immunoreactivity was observed in the proximal tubule. In the distal tubules of *B. f. japonicus* and the two aquatic species, the apical membrane was immunoreactive. In the connecting and the collecting tubules, the apical membrane of intercalated cells was immunoreactive in all species examined (*Figure 4B*). In the renal corpuscle, the neck segments and the intermediate segments, no immunoreactivity was observed in all species examined.

The functions of vacuolar (V-type) H^+-ATPase in various species and in various organs in mammals have been reviewed (e.g. Harvey, B.J., 1992). The major function of H^+-ATPase is acidifying the interior of intra-cellular organelles. The H^+-ATPase is also expressed specially on the plasma membrane of various ion-transporting epithelia or acid-secreting cells. The distribution of H^+-ATPase in the various nephron segments has been studied by both biochemical and immunohistochemical techniques (Brown *et al.*, 1988; Gluck and Nelson, 1992). Enzyme activities were high in the connecting tubule, intermediate in the distal convoluted tubule, and low in other segments of the rabbit nephron. However, there were significant species differences in H^+-ATPase activity along the nephron in mammals (Brown *et al.*, 1988; Gluck and Nelson, 1992). In the kidney of sub-mammalian vertebrates, the distribution of H^+-ATPase in the nephron has rarely been studied. The extra-renal structures, such as the bladder and the gills, have been more intensively studied in non-mammalian vertebrates (Braun and Dantzler, 1997). These organs might be more important in the elimination of H^+ and maintenance of acid–base balance. In amphibians, urinary bladder and skin have also been identified as sites of a plasma membrane H^+-ATPase activity (Brown *et al.*, 1987; Harvey, W.R., 1992; Ehrenfeld, 1998). According to an immunocytochemical study, H^+ secretion in the frog skin must be mediated by an H^+-ATPase located in the apical membrane of the MRCs (Ehrenfeld and Klein, 1997). The apical membrane of MRCs of frog skin is labelled by both H^+-ATPase antibodies and band-3 antibodies (Cl^-/HCO_3^- antiporter), and only a single type of functional MRC is involved in H^+ and base secretion in frog skin. However, a different type of MRC may exist in the toad skin (Budtz *et al.*, 1995). In the frog skin, therefore, active transepithelial H^+ secretion is generated by an H^+-ATPase pump localized in apical membranes of MRCs, whereas Na^+ absorption occurs through an amiloride-sensitive pathway in principal cells. In the present study, MRCs (or intercalated cells) were observed in late distal tubules of terrestrial and amphibious anurans. According to immunohistochemical observations, these MRCs were highly immunoreactive with H^+-ATPase antiserum. In mammals, two types of intercalated cells were defined in the collecting duct, which plays a major role in acid–base balance. Intercalated α-cells have a basolateral band 3-like anion exchanger and secrete H^+, while β-cells bind peanut agglutinin (PNA) apically and secrete HCO_3^- (Schuster *et al.*, 1991). H^+-ATPase is present in the luminal membrane of α-intercalated cells of the late distal nephrons (Brown *et al.*, 1988; Gluck and Nelson, 1992). The cells mediate the secretion of H^+ into the tubular lumen and tubular acidification is driven by a luminal H^+-ATPase. In amphibians, according to physiological studies, acidification of the tubular fluid occurs along the distal tubule. In the present study, MRCs being immunoreactive to H^+-ATPase antibody, were observed in the late distal tubules of the kidney in terrestrial and amphibious anurans. Apical membranes of the intercalated cells showed similar immunoreactivity in the aquatic species (*Xenopus laevis*). However, the mucus in the canalicular lumen of the flask cells, another cell type of the late distal segment, did not react with H^+-ATPase and PNA, in this study or other lectin-binding studies (Jonas *et al.*, 1991). Thus, the cell might not be related to acid–base balance in the late distal segments of the *X. laevis* kidney.

6. H^+,K^+-ATPase

6.1 *Immunohistochemical studies of H^+,K^+-ATPase*

Immunohistochemical observations and diagrammatic representations are shown in *Table 2C* and *Figures 4C* and *5*. In terrestrial and semi-aquatic species, H^+,K^+-ATPase was observed along the intercalated cells of the distal segments from early distal to collecting tubules. In *X. laevis*, H^+,K^+-ATPase immunoreactivity was observed along the proximal, distal, connecting and collecting tubules. H^+,K^+-ATPase immunoreactivity was predominantly localized in the apical membranes of cells in the collecting tubules (*Figure 4C1 and 4C3*). On the other hand, it is also observed in the basal membranes of cells in the proximal and distal tubules (*Figure 4C*). In the renal corpuscle, the neck segments and the intermediate segments, no immunoreactivity was observed in all species examined.

Two proton-transporting ATPases participate in active proton transport in the nephron, the electrogenic vacuolar H^+-ATPase and the electroneutral H^+,K^+-ATPase. The H^+,K^+-ATPases comprise a group of integral membrane proteins that belong to the K^+-ATPase subfamily or P-type cation-transporting ATPases. In mammals, various K^+-ATPase activities are present along the renal tubule. In addition to the Na^+,K^+-ATPase, three other K^+-ATPases are expressed in the rat nephron (Doucet, 1997). Type I is expressed in the collecting duct, type II in the proximal tubule and the loop of Henle in normal rats; type III is expressed in the collecting duct of chronically hypokalemic rats. Type III, being similar to the colonic H^+,K^+-ATPases, shares some pharmacological properties with Na^+,K^+-ATPase. The colonic H^+,K^+-ATPases and Na^+,K^+-ATPases are ouabain-sensitive and insensitive to Sch 28080; the enzymes show high molecular homology (Jaisser *et al.*, 1993). The colonic H^+,K^+-ATPase may transport Na^+, acting as a (Na^+,K^+) K^+-ATPase (Cougnon *et al.*, 1998). It has been established that proton secretion by the collecting duct of mammals is mediated, in part, by an H^+,K^+-ATPase, which is a membrane protein that extrudes H^+ and absorbs K^+ at the expense of ATP hydrolysis (Wingo and Cain, 1993). This renal transporter is functionally similar to the H^+,K^+-ATPases found in gastric parietal cells and colonic epithelia. Sch-28080-sensitive, ouabain-insensitive H^+,K^+-ATPase activity has been identified in the connecting tubule and the collecting ducts of rat and rabbit kidneys. In addition, an ouabain-sensitive isoform of H^+,K^+-ATPase has been observed in the kidneys of the latter species (Wingo and Smolka, 1995).

On renal H^+,K^+-ATPases of species other than mammalian, so far, little information has been obtained. Antibodies directed against a hog gastric H^+,K^+-ATPase α-subunit stained specific subdivisions of the neck, early and late proximal tubule, late intermediate tubules, both segments of the distal tubule and the early collecting duct of the renal tubule in the kidney of the elasmobranch spiny dogfish shark, *Squalus acanthias* (Swenson *et al.*, 1994). According to biochemical observations, however, the gastric H^+,K^+-ATPase-like pump is present in the late distal tubule but not in the proximal and early distal tubule (Planelles *et al.*, 1991). The H^+,K^+-ATPase may be involved in urinary acidification in the late distal tubule of amphibians. In non-mammalian tetrapods and teleosts, both net reabsorption and net secretion of potassium have been observed (see review, Braun and Dantzler, 1997). In amphibians, net secretion occurs along the late distal tubule and along the early distal tubule (Oberleithner *et al.*, 1983). Our data show that H^+,K^+-ATPase is located in the apical membranes of cells in collecting tubules. This suggests H^+,K^+-ATPase participates in both H^+ and K^+ transport in the distal nephron of amphibians as in mammals. On the other hand, immunoreactivity was also observed at the basal membranes of cells in the proximal and distal tubules. The polyclonal antibody to the H^+,K^+-ATPase-α subunit used in the present study may cross-react with renal Na^+,K^+-ATPase isoforms. Information on the intra-renal localization of the colonic H^+,K^+-ATPase is scant in lower vertebrates. The colonic H^+,K^+-ATPase may transport Na^+, acting as a (Na^+,K^+) K^+-ATPase (Cougnon *et al.*, 1998). It is necessary to examine the cellular and subcellular localization of the colonic H^+,K^+-ATPase in amphibians. The search for a function of the H^+,K^+-ATPase in the nephron segments in amphibians is warranted by our observations.

7. Aquaporins

7.1 *Immunohistochemical studies of aquaporins*

In lower vertebrates, AQP homologues were cloned from urinary bladder of anuran amphibians (Abrami *et al.*, 1994; Siner *et al.*, 1996; Ma *et al.*, 1996) and kidney of quail, *Coturnix coturnix.* (Nishimura et al., 2001; see also the contribution by Nishimura and Fan in this volume). We examined the kidney of the tree frog, *H. japonica,* using polyclonal antisera to mammalian aquaporins (AQP2 and AQP3). AQP2 immunoreactivity was observed in the apical membranes of the cells of the distal tubules, connecting tubules and the collecting ducts (data not shown). No

AQP3 immunoreactivity was observed in any segment of the nephron of the tree frog. Immunohistochemistry using anti-rat AQP2 antibody gave a similar result on the apical membrane staining of the collecting duct of birds, and increased in water-deprived and vasotocin-treated birds (Nishimura *et al.*, 2001).

Far more studies on amphibian species are required focusing on the cellular and sub-cellular localization of the AQPs and their function in nephron segments, as amphibian species may contribute important models to comparative physiology of water handling.

8. Summary

Amphibians inhabit areas ranging from completely aqueous to terrestrial environments and move between water and land. The kidneys of all anurans are similar at the gross morphological level: the structure of their nephrons is related to habitat. According to the observation by light and electron microscopy, the cells that make up the nephron differ among species. Immunohistochemical studies using antibodies to various ATPases showed a significant species difference depending on habitat. The immunoreactivity for Na^+,K^+ -ATPase was low in the proximal tubules but high in the basolateral membranes of early distal tubules to collecting ducts in all species. In the proximal tubule, apical membranes of the cells were slightly immunoreactive to H^+-ATPase antibody in aquatic species. In the connecting tubule and the collecting duct, the apical membrane of intercalated cells was immunoreactive in all species. In aquatic species, H^+,K^+-ATPase immunoreactivity was observed in cells along the proximal, distal tubule to the collecting duct. However, H^+,K^+-ATPase was present along the intercalated cells of the distal segments from early distal to collecting tubules in terrestrial and semi-aquatic species. In the renal corpuscle, the neck segment and the intermediate segment, immunoreactivities to ion pumps were not observed in any of the species examined. Taking together our observations, we conclude that in the aquatic species, a large volume of plasma must be filtered in a large glomerulus and the ultrafiltrate components are reabsorbed along a large and long proximal segment of the nephron. Control of tubular transport may be poorly developed when a small short distal segment of the nephron is observed. On the contrary, terrestrial species have a long and well-developed distal segment and regulation mechanisms of tubular transport may have evolved in these segments. Thus, the development of the late distal segments of the nephron is one of the important factors for the terrestrial adaptation.

Acknowledgements

We acknowledge the excellent technical assistance provided by M.Sc. M. Kamijo. We also thank Prof. K. Yamauchi, Division of Marine Biosciences, Graduate School of Fisheries Science, Hokkaido University and Prof. Y. Moriyama, Department of Biochemistry, Faculty of Pharmaceutical Sciences, Okayama University for their kind gifts of the rabbit antibodies against Na^+,K^+-ATPase and H^+-ATPase, respectively. This investigation was supported in part by funds of the Institute of Space & Astronautical Science, the Ministry of Education, Science, Sports and Culture of Japan, for Basic Experiments Oriented to Space Station Utilization.

References

Abrami, L., Simon, M., Rousselet, G., Berthonaud, V., Buhler, J.-M. and Ripoche, P. (1994) Sequence and functional expression of an amphibian water channel, FA-CHIP: a new member of the MIP family. *Biochim. Biophys. Acta* **1192:** 147–151.

Aperia, A., Fryckstedt, J., Holtback, U., Belusa, R., Cheng, X.J., Eklof, A.C., Li, D., Wang, Z.W. and Ohtomo, Y. (1996) Cellular mechanisms for bi-directional regulation of tubular sodium reabsorption. *Kidney Int.* **49:** 1743–1747.

Baines, A.D., Ho, P. and Drangova, R. (1992) Proximal tubular dopamine production regulates basolateral Na-K-ATPase. *Am. J. Physiol.* **262:** F566–F571.

Bankir, L. and de Rouffignac, C. (1985) Urinary concentrating ability: insights from comparative anatomy. *Am. J. Physiol.* **249:** R643–R666.

Beltowski, J., Gorny, D. and Marciniak, A. (1998) The mechanism of Na^+,K^+-ATPase inhibition by atrial natriuretic factor in rat renal medulla. *J. Physiol. Pharmacol.* **49:** 271–283.

Bentley, P.J. (1966) The physiology of the urinary bladder of Amphibia. *Biol. Rev.* **41:** 275–316.

Bentley, P.J. (1971) *Endocrines and Osmoregulation: A Comparative Account of the Regulation of Salt and Water in Vertebrates.* Springer, New York.

Bentley, P.J. and Yorio, T. (1979) Do frogs drink? *J. Exp. Biol.* **79:** 41–46.

Beuchat, C.A. (1996) Structure and concentrating ability of the mammalian kidney: correlations with habitat. *Am. J. Physiol.* **40:** R157–R179.

Boutilier, R.G., Stiffler, D.F. and Toews, D.P. (1992) Exchange of respiratory gases, ions, and water in amphibious and aquatic amphibians. In: *Environmental Physiology of the Amphibians* (eds M.E. Feder and W.W. Burggren). The Univ. Chicago Press, Chicago, IL, pp. 81–124.

Braun, E.J. and Dantzler W.H. (1997) Vertebrate renal system. In: *Handbook of Physiology. Section 13; Comparative Physiology*, Vol. 1 (ed. W.H. Dantzler). Oxford Univ. Press, Oxford, pp. 481–576.

Brown, D., Gluck, S. and Hartwig, J. (1987) Structure of the novel membrane-coating material in proton-secreting epithelial cells and identification as an H^+-ATPase. *J. Cell Biol.* **105:** 1637–1648.

Brown, D., Hirsch, S. and Gluck, S. (1988) Localization of a proton-pumping ATPase in rat kidney. *J. Clin. Invest.* **82:** 2114–2126.

Budtz, P.E., Christofersen, B.C., Johansen, J.S., Spies, I. and Willumsen, N.J. (1995) Tissue kinetics, ion transport and recruitment of mitochondria-rich cells in the skin of the toad (*Bufo bufo*) in response to exposure to distilled water. *Cell Tissue Res.* **280:** 65–75.

Buhagiar, K.A., Hansen, P.S., Gray, D.F., Mihailidou, A.S. and Rasmussen, H.H. (1999) Angiotensin regulates the selectivity of the Na^+/K^+-pump for intracellular Na^+. *Am. J. Physiol.* **277:** C461–C468.

Christensen, C.U. (1974) Effect of arterial perfusion on net water flux and active sodium transport across the isolated skin of *Bufo bufo bufo* (L.) *J. Comp. Physiol.* **93:** 93–104.

Cougnon, M., Bouyer, P., Planelles, G. and Jaisser, F. (1998) Does the colonic H, K-ATPase also act as an Na, K-ATPase? *Proc. Natl Acad. Sci. USA* **95:** 6516–6520.

Dantzler, W.H. (1989) *Comparative Physiology of the Vertebrate Kidney.* Springer, Berlin.

Doucet, A. (1997) H+, K+-ATPASE in the kidney: localization and function in the nephron. *Exp. Nephrol.* **5:** 271–276.

Doucet, A., Katz, A.I. and Morel, F. (1979) Na, K-ATPase activity along the rabbit, rat, and mouse nephron. *Am. J. Physiol.* **237:** F114–F120.

Ehrenfeld, J. (1998) Active proton and urea transport by amphibian skin. *Comp. Biochem. Physiol.* **119:** 35–45.

Ehrenfeld, J. and Klein, U. (1997) The key role of the H^+ V-ATPase in acid-base balance and Na^+-transport processes in frog skin. *J. Exp. Biol.* **200:** 247–256.

Feraill, E. and Doucet, A. (2001) Sodium-potassium-adenosinetriphosphatase dependent sodium transport in the kidney: hormonal control. *Physiol. Rev.* **81:** 345–418.

Fox, H. (1987) The skin of Amphibia. In: *Biology of the Integument. Vol. 2, Vertebrate* (eds K. Bereiter-Hahn, A.G. Matoltsy and K.S. Richards). Springer, Heidelberg, pp. 78–135.

Gluck, S. and Nelson, R. (1992) The role of the V-ATPase in renal epithelial H^+ transport. *J. Exp. Biol.* **172:** 205–218.

Guggino, W.B., Oberleitner, H. and Giebisch, G. (1988) The amphibian diluting segment. *Am. J. Physiol.* **254:** F615–F627.

Harvey, B.J. (1992) Energization of sodium absorption by the H^+-ATPase pump in mitochondria-rich cells of frog skin. *J. Exp. Biol.* **172:** 289–309.

Harvey, W.R. (1992) Physiology of V-ATPases. *J. Exp. Biol.* **172:** 1–17.

Hayashi, M., Yamada, J., Mitamura, T., Horii, T., Yamamoto, A. and Moriyama, Y. (2000) Vacuolar H^+-ATPase localized in plasma membranes of Malaria parasite cells, *Plasmodium falciparum*, is involved in regional acidification of parasitized erythrocytes. *J. Biol. Chem.* **275:** 34353–34358.

Hersey, R.M., Gattone, V.H. and Weisz, J. (1985) [Na-K]ATPase activity in proximal and distal tubules of the rat kidney: modification and application of a quantitative cytochemical technique. *Cell Biochem. Func.* **3:** 255–265.

Hickman, C.P., Jr and Trump, B.F. (1969) The kidney. In: *Fish Physiology. Vol. 1, Excretion, Ion Regulation, and Metabolism* (eds W.S. Hoar and D.J. Randoll). Academic Press, New York, pp. 1063–1096.

Jaisser, F., Coutry, N., Farman, N., Binder, H.J. and Rossier, B.C. (1993) Aputative H-K-ATPase is selectively expressed in surface epithelial cells of rat distal colon. *Am. J. Physiol.* **265:** C1080–C1089.

Jensen, L.J., Sorensen, J.N., Larsen, E.H. and Willumsen, N.J. (1997) Proton pump activity of mitochondria-rich cells. The interpretation of external proton-concentration gradients. *J. Gen. Physiol.* **109:** 73–91.

Jonas, L., Ostwald, C. and Walzel, H. (1991) Binding studies of gold labeled lectins on carbohydrate compounds of the flask cells in claw-frog kidney. *Acta Histochem.* **90:** 121–126.

Larsen, E.H. (1991) Chloride transport by high-resistance heterocellular epithelia. *Physiol. Rev.*: 235–288.

Long, W.S. (1973) Renal handling of urea in *Rana catesbeiana*. *Am. J. Physiol.* **224:** 482–490.

Ma, T., Yang, B. and Verkman, A.S. (1996) cDNA cloning of a functional water channel from toad urinary bladder epithelium. *Am. J. Physiol.* **271:** C1699–C1704.

Meister, B. and Aperia, A. (1993) Molecular mechanisms involved in catecholamine regulation of sodium transport. *Semin Nephrol.* **13:** 41–49.

Mills, J.W., Ernst, S.A. and DiBona, D.R. (1977) Localization of Na^+-pump sites in frog skin. *J. Cell Biol.* **73:** 88–110.

Mobjerg, N., Larsen, E.H. and Jespersen, A. (1998) Morphology of the nephron in the mesonephros of *Bufo bufo* (Amphibia, Anura, Bufonidae). *Acta Zool. (Stockh)* **79:** 31–51.

Mullen, T.L. and Alvarado, R.H. (1976) Osmotic and ionic regulation in amphibians. *Physiol. Zool.* **49:** 11–23.

Muller, J. and Kachadorian, W.A. (1984) Aggregate-carrying membranes during ADH stimulation and washout in toad bladder. *Am. J. Physiol.* **247:** C90–C98.

Nishimura, H., Fan, Z. and Zhang, L. (2001) Renal handling of water in vertebrates. In: *Perspective in Comparative Endocrinology: Unity and Diversity* (eds H.J.Th. Goos, R.K. Rastogi, H. Vaudry and R. Pierantoni). Monduzzi Editore, International Proceedings Division, Bologna, pp. 733–743.

Oberleithner, H., Lang, F., Greger, R., Wang, W. and Giebisch, G. (1983) Effect of luminal potassium on cellular sodium activity in the early distal tubule of *Amphiuma* kidney. *Pflugers Arch.* **396:** 34–40.

Ogawa, M. (1961) Comparative study on the external shape of the teleostean kidney with relation to phylogeny. *Sci. Rep. Tokyo Kyouiku Daigaku B* **10:** 61–68.

Parsons, R.H. (1994) Effects of skin circulation on water exchange. In: *Amphibian Biology. Vol.1 The Integument.* (eds H. Heatwole and G.T. Barthalmus). Surrey Beauty & Sons, Chipping Norton, NSW, pp. 132–146.

Planelles, G., Anagnostopoulos, T., Cheval, L. and Doucet, A. (1991) Biochemical and functional characterization of H^+, K^+-ATPase in distal amphibian nephron. *Am. J. Physiol.* **260:** F806–F812.

Puoti, A., May, A., Rossier, B.C. and Horisberger, J.D. (1997) Novel isoforms of the beta and gamma subunit of the Xenopus epithelial Na channel provide information about the amiloride binding site and extracellular sodium sensing. *Proc. Natl Acad. Sci. USA* **94:** 5949–5954.

Schuster, V.L., Fejes-Toth, G., Naray-Fejes-Toth, A. and Gluck, S. (1991) Colocalization of H^+-ATPase and band 3 anion exchanger in rabbit collecting duct intercalated cells. *Am. J. Physiol.* **260:** F506–F517.

Shoemaker, V.H. (1964) The effects of dehydration on electrolyte concentrations in a toad, *Bufo marinus. Comp. Biochem. Physiol.* **13:** 261–271.

Shoemaker, V.H. and Nagy, K.A. (1977) Osmoregulation in amphibians and reptiles. *Ann. Rev. Physiol.* **39:** 449–471.

Shoemaker, V.H., Hillman, S.S., Hillyard, S.D., Jackson, D.C., McClanahan, L.L. and Withers, P.C. (1992) Exchange of water, ions, and respiratory gases in terrestrial amphibians. In: *Environmental Physiology of the Amphibians* (eds M.E. Feder and W.W. Burggren). The Univ. Chicago Press, Chicago, IL, pp. 125–150.

Siner, J., Paredes, A., Hosselet, C., Hammond, T., Strange, K. and Harris, H.W. (1996) Cloning of an aquaporin homologue present in water channel containing endosomes of toad urinary bladder. *Am. J. Physiol.* **270:** C372–C381.

Stoner, L.C. (1985) The movement of solutes and water across the vertebrate distal nephron. *Ren. Physiol.* **8:** 237–248.

Sweeney, G. and Klip, A. (1998) Regulation of the Na^+/ K^+-ATPase by insulin: why and how? *Mol. Cell Biochem.* **182:** 121–133.

Swenson, E.R, Fine, A.D., Maren, T.H., Reale, E., Lacy, E.R. and Smolka, A.J. (1994) Physiological and immunocytochemical evidence for a putative H, K-ATPase in elasmobranch renal acid secretion. *Am. J. Physiol.* **267:** F639–F645.

Tanii, H., Hasegawa, T., Hirakawa, N., Suzuki, M. and Tanaka, S. (2000) Cloning of two distinct cDNAs encoding water channel proteins from ventral skin of the Japanese tree-frog, *Hyla japonica. Proc. Jap. Soc. Comp. Endocrinol.* Abst. of 25th Meeting **15:** 67.

Uchiyama, M., Murakami, T., Yoshizawa, H. and Wakasugi, C. (1990) Structure of the kidney in the crab-eating frog, *Rana cancrivora. J. Morphol.* **204:** 147–156.

Ura, K., Soyano, K., Omoto, N., Adachi, S. and Yamauchi, K. (1996) Localization of Na^+,K^+-ATPase in tissues of rabbit and teleosts using an antiserum directed against a partial sequence of the α-subunit. *Zool. Sci.* **13:** 219–227.

Verrey, F., Beron, J. and Spindler, B. (1996) Corticosteroid regulation of renal Na,K-ATPase. *Miner. Electrolyte Metab.* **22:** 279–292.

Welling, P.A., Caplan, M., Sutters, M. and Giebisch, G. (1993) Aldosterone-mediated Na/K-ATPase expression is alpha 1 isoform specific in the renal cortical collecting duct. *J. Biol. Chem.* **268:** 23469–23476.

Wilson, P.D. and Horster, M.F. (1983) Differential response to hormones of defined distal nephron epithelia in culture. *Am. J. Physiol.* **244:** C166–C174.

Wingo, C.S. and Cain, B.D. (1993) H^+,K^+-ATPase The renal H^+,K^+-ATPase; physiological significance and role in potassium homeostasis. *Ann. Rev. Physiol.* **55:** 323–347.

Wingo, C.S. and Smolka, A.J. (1995) Function and structure of H-K-ATPase in the kidney. *Am. J. Physiol.* **269:** F1–F16.

9

Sodium and water transport and urine concentration in avian kidney

H. Nishimura and Z. Fan

1. Introduction

Reptiles, birds and mammals share a common evolutionary origin, but the reptilian evolutionary line leading to the mammals (*Synapsida*) departed from the reptilian–avian line (archosauria) at a very early stage of tetrapod evolution (King and King, 1979; Young, 1981). Avian kidneys have several unique morphological and functional features (Dantzler and Braun, 1980; Braun, 1993; Nishimura, 1993). Unlike other vertebrate kidneys, avian and mammalian kidneys show two distinct zonations, the cortical and medullary zones. Avian kidneys have short cortical loop-less (reptilian-type) nephrons and longer looped (mammalian-type) nephrons that are situated deeper in the medullary zone. Essential epithelial transport systems for ions and water are present in bird kidneys; the architectural structure of the avian nephron and vascular system, however, is simpler than in mammals, as is the organizational relationship between nephrons and blood vessels (Nishimura, 1993; Braun and Dantzler, 1997).

Birds are the only non-mammalian vertebrates that can produce hyper-osmotic urine by a countercurrent urine concentration mechanism utilizing recycling of a single solute (NaCl) (Nishimura, 1989; Nishimura, 1993); thus, they adapt to terrestrial life by water conservation. The corticomedullary osmotic gradient in birds is, however, primarily composed of NaCl with no complement by a nitrogenous compound such as urea. Uric acid, the major nitrogen metabolite in birds, has low solubility and thus does not contribute to the formation of the osmotic gradient along the renal medulla. Furthermore, the loop-less, reptilian-type nephrons that constitute the majority of avian nephrons produce only hypo-osmotic, or at most iso-osmotic, urine. Hence, the medullary collecting duct (CD) flow and urine concentration are significantly influenced by the drainage of dilute cortical urine. In birds, as in mammals, an aquaporin2 (AQP2) water channel is present at the apical membrane of the cortical and medullary CD (Nishimura *et al.*, 2000; Nishimura *et al.*, 2001;

Osmoregulation and Drinking in Vertebrates, edited by N. Hazon & G. Flik.
© 2002 BIOS Scientific Publishers Ltd, Oxford

Zhang *et al.*, 2001; Cui *et al.*, 2001); it is regulated by arginine vasotocin (AVT), an avian antidiuretic hormone (ADH). In this brief review, we intend to summarize the structural and functional basis of sodium and water transport and of urine concentration in birds, focusing on the loop of Henle. We will also discuss recent advancements in AQP water channel research in mammalian and non-mammalian kidneys. The handling of Na and water via renal glomeruli and proximal tubules and the transport of monovalent ions other than NaCl or divalent ions are beyond the scope of this review, and readers should refer to the recent comprehensive review by Braun and Dantzler (1997). Birds live in a variety of habitats, and the role of the kidney varies in accordance with adaptations to environments. Moreover, osmoregulatory organs other than the kidney (such as the gastrointestinal tract, cloaca and nasal gland) that complement or interact with kidneys play an important role in ion and water balance in birds, but these aspects will not be discussed in this chapter.

2. Functional morphology of the loop of Henle

2.1 *Architectural organization of the medullary cone*

The cortical loop-less nephron consists of a proximal tubule, intermediate segment, distal tubule, and collecting tubule and duct. The looped nephron, which resides in a deeper zone, has the loop of Henle instead of the intermediate segment. Although it has not been established that the intermediate segment is an evolutionary precedent of Henle's loop, the intermediate segment adjacent to the distal tubule of Japanese quail, *Coturnix coturnix*, shows an ouabain-sensitive, lumen-positive transepithelial voltage similar to that of the thick ascending limb of the looped nephron (Nishimura *et al.*, 1986). In looped nephrons of avian kidneys, the straight segment of the proximal tubule runs into the narrower descending limb (DL) rather abruptly. The diameter of the DL (10–20 μm) gradually increases and moves into the thick limb (TL). This transition from DL to TL occurs before the hairpin turn; thus, bird kidneys, differing from mammalian kidneys with their long-looped nephrons, lack the thin ascending limb and inner medulla (Johnson and Mugaas, 1970). The thick ascending limb moves into the distal tubule, collecting tubules, and CDs; the CDs run parallel to Henle's loop, indicating that looped nephrons have a structural basis for producing concentrated urine by a countercurrent multiplier mechanism (*Figure 1*). Furthermore, the hairpin turns occur in a graded fashion at various levels of the medullary cone (Nishimura *et al.*, 1989; Casotti *et al.*, 2000). Cross sections of the medullary cones of some species such as the *Passeriformes* (Johnson and Mugaas, 1970) or *Coturnix* quail (Nishimura *et al.*, 1989) reveal a highly organized pattern that includes a central core comprising the thin DL and capillaries surrounded by a ring of CDs. The thin DLs are often in close contact with the CDs, whereas the TLs are distributed both inside and outside of the core. In contrast, all tubular elements are intermingled, with no specific organization in the medullary cones of non-passerines and other species. The functional significance of this architectural organization and its relation to urine-concentrating capacity remains to be determined.

2.2 *Thin descending limbs and heterogeneity*

The DL of avian looped nephrons has thin walls (~20 μm diameter) and also shows heterogeneity of epithelial cells. Three types of cells (A, B and C) have been identi-

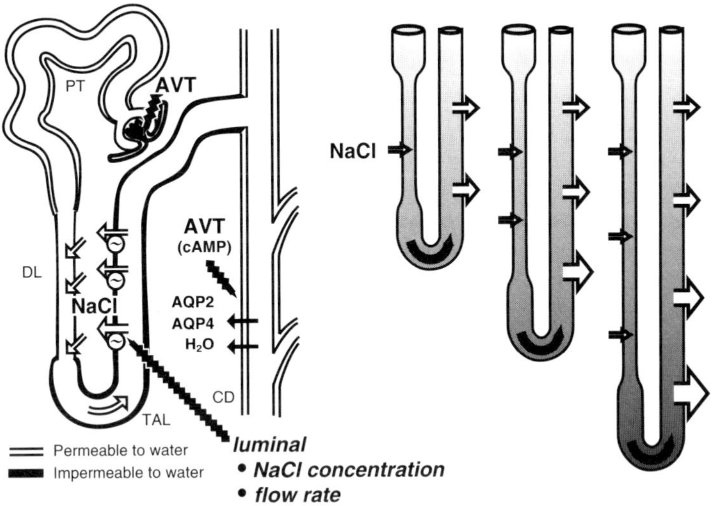

Figure 1. Left. *A proposed model for urine concentration in the mammalian-type nephron of* Coturnix *quail, showing transport properties of NaCl and water in various nephron segments. NaCl is actively extruded from the thick ascending limb (TAL) and enters the descending limbs (DLs) by simple diffusion without the osmotic accompaniment of water; the NaCl is recirculated for maximum operation of a single-effect countercurrent multiplier system. Active NaCl transport in the TAL appears to be enhanced by perfusion flow rate or luminal NaCl concentration. Immunohistochemical and molecular studies indicate that AQP2 and AQP4 are present in the CD, regulating water transport across the CDs.* ***Right.*** *Hypothetical cascade transport of NaCl in medullary cone, showing that more NaCl extrusion occurs near the tip of the cone, which may help to enhance an osmotic gradient (reproduced with modification from Osono and Nishimura, 1994). PT: proximal tubule.*

fied (Braun and Reimer, 1988; Nishimura *et al.*, 1989). Type A cells, located near the proximal tubule–DL junction, are relatively tall and have numerous apical microvilli, a moderate number of mitochondria, and lateral interdigitation with neighbouring cells, forming a labyrinthine structure; whereas Type C cells are flat, with little interdigitation, and are seen adjacent to the DL–TL transition (Nishimura *et al.*, 1989). Type B cells are an intermediate form between A- and C-type cells. Differing from the case in mammalian DL cells, basal infoldings are scarce in the three types of avian DL cells. The length of the thin DL of Henle's loop correlates with the entire length of the loop, whereas the length of the TL preceding the hairpin turn varies (Casotti *et al.*, 2000).

2.3 *Thick limbs, collecting tubules and collecting ducts*

The TLs of Henle's loop in avian and mammalian kidneys and the diluting segments of freshwater (FW) teleosts and aquatic amphibians show similar morphological characteristics (Kriz *et al.*, 1980; Braun and Reimer, 1988; Nishimura *et al.*, 1983; Nishimura *et al.*, 1986). The epithelial cells of TLs are cuboidal and have abundant mitochondria between the basal infoldings situated perpendicular to the base. The

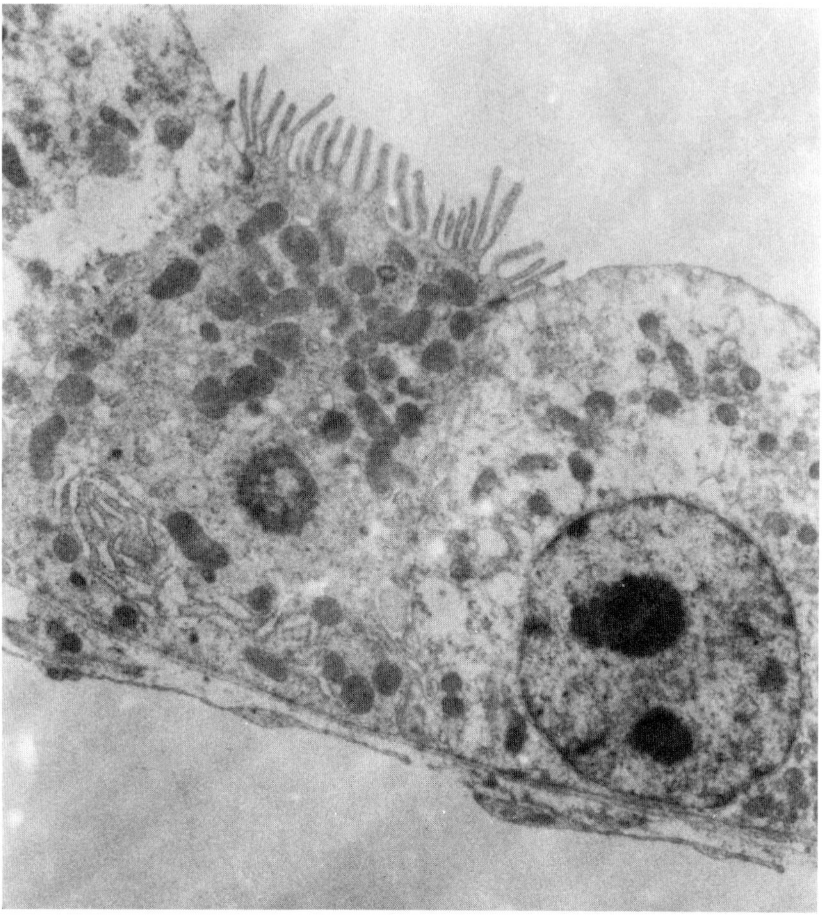

Figure 2. *A representative transmission electron micrograph of a dark cell (intercalated cell) and a mucus-secreting light cell in the cortical collecting duct from* Coturnix *quail. The dark cell contains abundant mitochondria, microvilli and vesicles on the apical side.*

CD cells in the cortical zone contain dark electron-dense cells that have apical microvilli and vesicles, resembling intercalated cells, and light mucus-secreting cells that have large vacuoles, particularly in the apical side of the nucleus (Nicholson and Kendall, 1983; Nishimura *et al.*, 1996) (*Figure 2*). Abundant interdigitation and tight deep junctions are seen between mucus-secreting cells and also between intercalated cells and mucus-secreting cells.

3. NaCl and water transport in the loop of Henle

3.1 *Descending limb*

The epithelial cells of the DL also exhibit functional heterogeneity (Nishimura *et al.*, 1989). The upper segment of the DL (A cells) has no net volume flux, has low diffusional and osmotic water permeability, and has zero transepithelial voltage. Type A cells have high and nearly equal permeability to Na and Cl ions, suggesting that

NaCl is likely to enter the DL unaccompanied by water. Mammalian kidneys show structural and functional heterogeneity of thin limbs within nephrons, between short-looped and long-looped nephrons, and among species (Kriz *et al.*, 1980; Bankir and de Rouffignac, 1985) (*Table 1*), and the epithelia of thin limbs are classified into four types (type I–VI) based on cell thickness, abundance of luminal microvilli, mitochondria and interdigitation, and the depth of tight junctions (Kriz *et al.*, 1980). Type I represents the DL of the short-looped nephron containing thin flat epithelia with little interdigitation and deep tight junctions. This segment shows relatively low permeability to Na^+ and Cl^- whereas its permeability to urea and water is high (Imai *et al.*, 1984). In rodents, the thin limbs of short-looped nephrons run inside the vascular core (Bankir and de Rouffignac, 1985), to which urea from the papilla is diffused from the ascending vasa recta via urea transporters (UTA2 and UTB1; for review, see Trinh-Trang-Tan and Bankir, 1998; Bankir and Trinh-Trang-Tan, 2000). The type II segment, the upper part of long-looped nephrons, consists of epithelial cells that have much interdigitation with neighbouring cells, shallow tight junctions, and high Na (cation-selective) and water permeability. The morphological and functional features of the type II segment are more clear in rodents than in humans or rabbits (Kriz *et al.*, 1980). The type III segment includes the lower part of long-looped nephrons and has a much simpler structure and lower Na permeability than the type II segment (Imai *et al*, 1988); whereas the type IV segment, the thin ascending limb of the loop of Henle, has moderate cellular interdigitation and permeability to Cl^- via chloride channels that reside at both the apical and basolateral side of epithelial cells, inducing facilitated diffusion (Uchida *et al.*, 1993). This segment has shallow tight junctions (Kriz *et al.*, 1980).

Although the heterogeneity of thin limbs appears less complex in avian nephrons, there are common features between the upper part of the DL of avian looped nephrons and that of mammalian long-looped nephrons. This similarity suggests that avian looped nephrons may evolutionarily correspond to the outer medullary part of mammalian long-looped nephrons rather than to the cortical short-looped nephrons. Major differences in transport properties, however, exist between the upper part of the DL of avian looped nephrons and that of mammalian long-looped nephrons: water permeability is low, while permeabilities to Na^+ and Cl^- are high and nearly equal in the former; whereas very high water permeability and cation selectivity are seen in the latter (*Table 1*). These differences in transport characteristics appear important for the difference between countercurrent urine-concentrating mechanisms in avian and mammalian kidneys.

3.2 *Thick ascending limb*

In looped nephrons of bird kidneys, the thick ascending limb (TAL) shows characteristics of diluting segments, including 1) lumen-positive transepithelial voltage, 2) furosemide- or bumetanide-inhibitable Na^+-$2Cl^-$-K^+ co-transport and an ouabain-sensitive basolateral Na^+-K^+ pump, and 3) low volume flux and osmotic permeability (for review, see Miwa and Nishimura, 1986; Nishimura, 1993; Braun and Dantzler, 1997) (*Table 2* and *Figure 3*). In the isolated perfused TAL of the quail kidney, increased perfusion flow rate linearly increases the lumen-to-bath Na^+ and Cl^- flux and the flux coefficient (index for NaCl permeability), whereas little change is noted in the bath-to-lumen flux or flux coefficient (Osono and Nishimura, 1994). Increased

Table 1. *Morphological and functional properties of thin descending limb of Henle in quail and hamster*

	Quail	Hamster		
	tDL_u	SDL	LDL_u	LDL_l
Epithelial cells				
Height	Relatively tall	Flat	Relatively tall	Flat
Microvilli	++	±	+++	±
Interdigitation	+++	±	+++	±
Mitochondria	++	±	+++	±
Basal infoldings	—	±	++	±
Junction	Tight, shallow	Tight, deep	Tight, shallow	Tight, deep
V_t (mV)	0	0	0	0
Diffusional ΔV_t (mV)	1.9 ± 0.2	2.9 ± 0.4[a]	– 9.4 ± 1.0[a]	1.3 ± 0.1
J_v, isosmotic bath ($nl\ min^{-1}\ mm^{-1}$)	0.21 ± 0.14	–0.20 ± 0.12	0.15 ± 0.05	–
J_v, hyper-osmotic bath ($nl\ min^{-1}\ mm^{-1}$)	0.49 ± 0.19[b]	14.65 ± 4.36[c]	22.60 ± 2.20[c]	29.67 ± 6.35[d]
L_p ($10^{-9}\ cm^2\ s^{-1}\ atm^{-1}$)	83.3 ± 32.4	1394 ± 506	2600 ± 328 1760[e]	1693 ± 517
P_{dw} ($10^{-7}\ cm^2\ s^{-1}$)	73.0 ± 7.8	–	–	–
P_{Na} ($10^{-7}\ cm^2\ s^{-1}$)	31.7 ± 2.3	2.9 ± 1.4	41.0 ± 5.4	2.4 ± 0.8
P_{Cl} ($10^{-7}\ cm^2\ s^{-1}$)	24.9 ± 3.56	0.9 ± 0.2	3.8 ± 0.6	–
P_{urea} ($10^{-7}\ cm^2\ s^{-1}$)	–	5.1 ± 1.4	1.4 ± 0.3	7.9 ± 4.1
Reference	Nishimura *et al.*, 1989	Imai *et al.*, 1984	Imai *et al.*, 1984	Imai *et al.*, 1988

Note: Osmolality units are mOsm $kg^{-1}\ H_2O$.
[a]From Imai *et al.*, 1984. [b]Perfusate 312, bath 364 with sucrose. [c]Perfusate 304, bath 405 with raffinose. [d]Perfusate 300, bath 477 with sucrose.
[e]From Imai *et al.*, 1988.
tDL_u = upper segment, thin descending limb. SDL = descending limb of short-looped nephron. LDL_u = upper segment of descending limb of long-looped nephron. LDL_l = lower segment of descending limb of long-looped nephron. V_t = transepithelial voltage. J_v = net volume flux.
L_p = osmotic water permeability. P_{dw} = diffusional water permeability. P_{Na}, P_{Cl}, P_{urea} = lumen-to-bath flux coefficent of Na^+, Cl^-, urea as an index for permeability.

Table 2. *Transport properties of diluting segments from representative species*

Species	Water flux		Net solute flux[a]		References
	J_v (nl mm^{-1} min^{-1} min^{-1})	L_p (10^{-9} cm^2 s^{-1} atm^{-1})	Na (pEq mm^{-1} s^{-1} min^{-1})	Cl	
Fish					
River lamprey, FW (*Lampetra fluviatilis*)					Logan *et al.*, 1980
Dogfish, SW (*Squalus acanthias*)					Friedman and Hebert, 1990; Hebert and Friedman, 1990
Rainbow trout (*Oncorhynchus mykiss*)	0.12 ± 0.11	13.8 ± 3.5	—	66.5 ± 20.6	Nishimura *et al.*, 1983
Amphibians					
Frog (*Rana pipiens*)	0.07 ± 0.09[b]	5.0	81.1 ± 10.1	56.6 ± 6.7	Stoner, 1977
Amphiuma sp.	0.07[b]	—	—	33.4 ± 1.4	Oberleithner *et al.*, 1982
Birds					
Quail (*Coturnix coturnix*)	–0.01 ± 0.04	12.8 ± 2.1	242 ± 49	271.8 ± 32.9	Miwa and Nishimura, 1986
Mammals					
Rabbit (*Oryctolagus cuniculus*)					
Cortex	0.13 ± 0.02[b]	5.0 ± 0.7	79.2 ± 8.4	59 ± 10	Burg and Green, 1973a, b
Medulla	0.02 ± 0.01[d]	0.9 ± 2.2	28 ± 8	19 ± 5	Rocha and Kokko, 1973
Mouse (*Mus musculus*)	–0.01 ± 0.04	7.5 ± 0.9	—	94 ± 8	Hall and Varney, 1980
					Hall and Varney, 1980

Note: [a]Net solute fluxes were measured with iso-osmotic bath and perfusate using concentrations (mEq l^{-1}) of Na (120–150) and Cl (114–151) in all species shown except frog (Na = 115, Cl = 95) and Congo eel (Na = 97, Cl = 84–100). [b] Measured with osmotic gradient. [c] Difference between efflux and influx. [d]Units of nl min^{-1}. J_v = net volume flux. L_p = osmotic water permeability. Dogfish, intermediate segment IV.

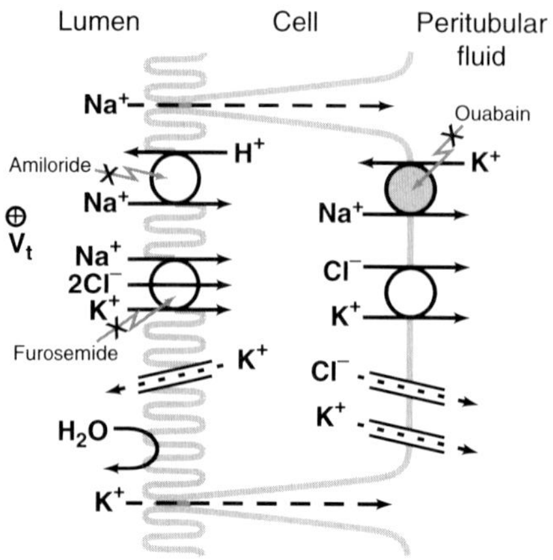

Figure 3. *Schematic presentation of epithelial ion and water transport properties of diluting segments (early distal tubules of teleosts and amphibians and thick ascending limb of bird). The luminal Na+-2Cl-K+ co-transporter and Na+-H+ exchanger are operated based on a steep Na+ gradient formed by an active basolateral Na+-K+-ATPase pump. Transepithelial voltage (Vt) is positive in the lumen. K+ and Cl- are moved into or out of cells by diffusion or via the K+ or Cl- channel. Data are from Dietl and Stanton, 1993; Nishimura et al., 1983; 1986; Miwa and Nishimura, 1986.*

NaCl concentration also increases NaCl fluxes, suggesting that increased luminal loading of NaCl evoked by increased volume flow or NaCl concentration facilitates NaCl extrusion (reabsorption), which may help enhance the imposition of the medullary osmotic gradient. The increase in NaCl delivery (increases in substrate) that facilitates active NaCl transport or the pressure-induced increase in para-cellular NaCl transport may account for this flow/concentration-dependent increase in NaCl transport in the TAL. In mammalian kidneys, vasopressin stimulates Cl transport (reabsorption) across the TAL (Hall and Varney, 1980; Hebert *et al.*, 1981), whereas AVT increases neither transepithelial voltage nor lumen-to-bath Cl transport in isolated and perfused TALs from quail 4–6 weeks old (Miwa and Nishimura, 1986). AVT markedly increases cAMP production in the TAL from the medulla of the house sparrow, *Passer domesticus* (Goldstein *et al.*, 1999).

4. Water transport in the collecting duct

The function of the CD is essential for water conservation. We found that in isolated perfused medullary CDs of *Coturnix* quail kidneys, the volume flux is nearly zero, but it tends to increase when an osmotic gradient is imposed (Nishimura *et al.*, 1996). When these birds are acutely treated with AVT, the volume flux increases only in water-deprived quail. Furthermore, the basal diffusional water permeability (tritiated water flux coefficient) in medullary CDs is considerable. The diffusional water permeability is increased slightly by AVT and only modestly by forskolin (FSK), which stimulates adenylate cyclase, whereas an FSK analogue that does not stimulate adenylate cyclase shows no effect (Nishimura *et al.*, 1996). The increase in diffusional water permeability caused by AVT in birds is much lower than the increase evoked by ADH in rat and hamster CDs (Kondo and Imai, 1987; de Rouffignac *et al.*, 1993). In medullary cones, cAMP levels are also increased only slightly by AVT but markedly by FSK, suggesting that the relatively low diffusional water permeability response to AVT is not due to insufficient adenylate cyclase activity. An AVT-sensitive aquaporin (AQP) water channel is present in CDs of medullary cones (see below).

5. Countercurrent multiplier mechanism for urine concentration

There are three basic requirements for the DL and TAL of the loop of Henle to act as a single-effect countercurrent multiplier: 1) a source of energy to create a transverse difference in osmotic pressure, 2) differences in epithelial permeability, and 3) countercurrent flow. First, the transport properties of the TAL of *Coturnix* quail agree with a luminal Na^+-dependent Cl^- transport driven by a basolateral Na^+-K^+-ATPase-operated pump without water accompaniment (Miwa and Nishimura, 1986). The TAL is thus able to provide an energy source. Second, the epithelial permeabilities of the DL and TAL differ. Third, the architectural organization of the medullary cone indicates that there is a counterflow between the descending and ascending limbs of Henle's loop. It is necessary, however, to prove the presence of counterflow by micropuncture. Thus, the *Coturnix* kidney satisfies the basic requirements mentioned above. The DL is highly and nearly equally permeable to Na and Cl with no active transport, whereas the osmotic and diffusional permeabilities to water are low (Nishimura *et al.*, 1989). Therefore, tubular urine may be concentrated by the addition of NaCl while it passes through the DL. In contrast, urine dilution occurs by active NaCl absorption in the TAL. Net NaCl flux in the TAL is further facilitated by increased luminal NaCl delivery evoked by increased flow rate or NaCl concentration (Osono and Nishimura, 1994). Therefore, in avian kidneys, urine concentration and dilution occur by a combination of passive diffusion and active transport, and the recycling of a single solute (NaCl). This single effect may be multiplied by counterflow and the graded elongation of hairpin structures, forming an osmotic gradient along the medullary cones (*Figure 1*). The low water permeability of the DL helps to maintain the bulk of volume flow from the proximal tubules, which in turn facilitates NaCl transport from the luminal to the interstitial side via the perfusion flow-dependent mechanism discussed above (Osono and Nishimura, 1994).

In rodent kidneys, DLs of cortical short-looped nephrons are wrapped by a vascular core consisting of ascending and descending vasa recta that forms a short circuit pathway for urea recirculation from the inner medulla to the cortical CDs (Bankir and de Rouffignac, 1985). In addition, ion and water exchange between the ascending and descending vasa recta efficiently move water out of the inner medulla and papilla, providing 'insulation' of these areas that helps to maintain the medullary osmotic gradient. The presence of such a mechanism has not been explored in avian kidneys. The transport characteristics of diluting segments reported by various groups are summarized in *Table 2*. All diluting segments show zero volume flux, low osmotic permeability, and the presence of a net NaCl flux. Net Na and Cl fluxes in the avian TAL are much higher than those in diluting segments/TALs of kidneys from other vertebrate animals. This may be causally related to the fact that the medulla of avian kidneys consumes more energy to operate a countercurrent multiplier system for urine concentration because the osmotic gradient along the medullary cone is solely maintained by NaCl without the contribution of urea. In newborn rabbits, in which the structure of the medulla has not completely matured, the medullary osmotic gradient is low (Horster, 1978).

In general, osmolal urine-to-plasma (U/P) ratios are lower in birds (Braun, 1980; Goldstein and Braun, 1989) than in mammals, particularly rodents. In most birds examined, the maximum osmolal U/P ratio is below 2.5 after dehydration. In avian kidneys, the hypo-osmotic urine that drains from the cortical loop-less nephrons to the medullary CDs decreases the final urine concentration. Considering the smaller number of looped nephrons, usually 15–30% (Braun, 1980; Braun and Dantzler, 1997), and the fact that only a limited number of nephrons in Henle's loop reach the tip of the medullary cone, the looped nephrons of avian kidneys must have an efficient urine-concentrating mechanism in order to double the urine osmolality. Furthermore, in birds, the plasma osmolality increases considerably during water deprivation; the higher plasma osmolality apparently reduces the osmolal U/P ratio.

6. Aquaporin water channels

6.1 *Overview*

Aquaporins (AQPs) are a family of small, hydrophobic proteins (major intrinsic proteins, MIPs) that were originally cloned in mammalian lens as MIP26 (for review, see Yamamoto and Sasaki, 1998; Borgnia *et al.*, 1999; Agre, 2000; Verkman and Mitra, 2000). Ten major mammalian AQPs and many isoforms have been identified. AQPs are expressed in a variety of epithelial transport tissues. Two AQP subgroups appear to be present: AQPs that are selectively permeable to water (AQPs 1, 2, 4, 5 and 6) and aquaglyceroporins (3, 7 and 9) that are permeable to both water and glycerol (Yamamoto and Sasaki, 1998; Verkman and Mitra, 2000). AQP8 is permeable to water and urea in some species. Aquaglyceroporins contain two additional peptide spans in the second and third external loops. AQPs are phylogenetically old molecules. In non-mammalian vertebrates, AQPs have been cloned from teleost fish (Cutler and Cramb, 2000; Virkki *et al.*, 2001), amphibians (Siner *et al.*, 1996; Abrami *et al.*, 1994), and birds (Nishimura *et al.*, 2001; Cui *et al.*, 2001). AQPs are present in insects, microbial organisms and plants (for review, see Chrispeels and Agre, 1994; Borgnia *et al.*, 1999).

AQP channels have a common molecular architecture, consisting of homotetramers in which each monomer (26–34 kDa in size) contains six membrane-spanning alpha-helical segments forming hydrophilic loops A to E (or loops 1–5) with the amino and carboxyl termini in the intra-cellular milieu (for review, see Borgnia *et al.*, 1999; Agre, 2000; Verkman and Mitra, 2000; Ashcroft, 2000). Each molecule reveals two internal repeats oriented at 180° to each other in the membrane: first repeat, transmembrane domains 1–3; and the second repeat, 4–6. Both repeats contain a highly conserved short sequence, the NPA motif (asparagine-proline-alanine). The two NPA motifs located in loops B and E are juxtaposed and form a single aqueous channel spanning the bilayer (hourglass model) to comprise a single water-selective channel. Mutation of this region reduces water permeability (Jung *et al.*, 1994). Study of its three-dimensional structure using cryoelectron microscopy techniques reveals four subunits; each has six membrane-spanning alpha helices, slightly tilted, and the centre of the monomers leads to the pore of the water channel (~8 A) (Mitra, 2001). The function of the AQP water channel protein is usually determined by measuring the osmotic water permeability of AQP-expressing *Xenopus* oocytes. After AQP cRNA is injected into the oocyte and expressed, the expansion of the oocyte volume in a hypo-osmotic medium is measured using an NIH Image program.

6.2 *Renal aquaporins*

Most cell membranes show some water permeability, either by simple diffusion via the lipid bilayer or via ion channels. Tissues that have high water permeability possess a water-selective channel. It has been shown that AQPs 1, 2, 3, 4, 6 and 7 are expressed in mammalian kidneys (Yamamoto and Sasaki, 1998). While AQP2 (and possibly AQP6) is inducible (see below), AQPs 1, 3 and 4 are constitutively expressed (for review, see King *et al.*, 2000; Yamamoto and Sasaki, 1998; Verkman and Mitra, 2000).

AQP1 is a channel-forming integral protein of 28 kDa (CHIP28, 276 amino acid residues in human; Preston and Agre, 1991) purified from red blood cells (co-purified with rhesus blood group antigen) and highly homologous to MIP26. Oocytes injected with mRNA encoding AQP1 swell rapidly and burst when placed in hypo-osmotic solution, indicating that AQP1 acts as a water channel. AQP1 is mercury-sensitive and is expressed widely in mammalian water-permeable tissues, including proximal tubules (microvilli of the apical brush border and less intensely in the basolateral membrane), the thin DL of Henle's loop, the DL of the vasa recta of the kidney (Nielsen *et al.*, 1995), gall bladder, red blood cells, eye, respiratory tract, and alveoli. AQP1 is also seen in endothelial cells of the descending vasa recta, mediating water transport across the vasa recta. AQP1 is selective to water, but not to urea, protons, hydroxyl ions or ammonium ions.

AQP2 (AQP-CD, 29 kDa, 271 amino acid residues in human) has been cloned from kidneys of rats (Fushimi *et al.*, 1993), humans, mice and other species (Yamamoto and Sasaki, 1998). AQP2 has 42% homology with CHIP28 and is selective to water, but not to urea. The AQP2-specific region (FDDHWVFW, 198–205) is the same for rat, human, mouse and sheep AQP2; but the overall amino acid sequence, including this coding region, differs significantly from that of other renal AQPs. AQP2 is selectively expressed in connecting tubules and cortical and medullary CDs and is localized at the apical membrane and subapical regions of the principal cells (Fushimi *et al.*, 1994). Vasopressin acutely stimulates trafficking of the AQP2 molecule from the subapical endosomes to the membrane (see 6.3). Furthermore, dehydration and treatment with vasopressin stimulate both AQP protein and mRNA, whereas their expression is inhibitable by V_1 antagonists (Hayashi *et al.*, 1994).

AQP3 shows modest permeability to urea and glycerol in addition to water and is expressed in the basolateral membrane of the cortical and outer medullary CDs and in the epithelial cells of the gastrointestinal tract, trachea and other tissues (Echevarria *et al.*, 1994). AQP3 is constitutively present in membrane without trafficking from vesicles, but a slight increase in mRNA and protein has been reported after long-term water deprivation (Yamamoto and Sasaki, 1998).

AQP4 has been cloned independently by two groups (for review, see Yamamoto and Sasaki, 1998) from the rat lung and brain cDNA library and is constitutively expressed in the basolateral side of the principal cells of the inner medulla (Terris *et al.*, 1995), bronchial epithelia, eye epithelia and other tissues. AQP4 is not regulated through trafficking between the intra-cellular vesicles and basolateral membrane. Water restriction does not alter the level of AQP4 expression, and AQP4 is mercury-insensitive. Phorbol ester inhibits AQP4 expression, suggesting that protein kinase C may be involved in cellular signalling (Han *et al.*, 1998).

AQP6 was isolated from rat and human kidney and its gene sequence locus is similar to that of AQP2, suggesting that their evolution may be closely related. AQP6 is expressed in vesicles of the epithelial cells of the proximal tubule and in intercalated cells of the CDs (Yasui *et al.*, 1999b) and is identified as an intra-cellular water channel (King *et al.*, 2000). The water permeability of AQP6-expressing membrane is relatively small. AQP6 may be important for regulation of acid–base balance (Yasui *et et al.*, 1999a). **AQP7** (Ishibashi *et al.*, 1997) is a glyceroAQP and is expressed abundantly in testis, heart and also in the brush border of the straight, but not convoluted, proximal tubules.

6.3 *Cellular mechanism and regulation*

The cellular mechanism of the AQP2 water channel has been extensively studied. In the principal cells of the CDs, AQP2 is primarily regulated by an AVP/cAMP-dependent vesicular trafficking mechanism, attributed to the so-called 'Shuttle hypothesis' (for review, see Laycock and Hanoune, 1998; Verkman and Mitra, 2000) via V_2 receptors. In the inactive state, AQPs are located in subapical vesicles and endosomes. Upon AVP binding to basolateral receptors and activation of the adenylate cyclase/cAMP and protein kinase A (PKA) system, vesicles containing AQP are translocated to the apical membrane, where AQPs are inserted by a mechanism similar to exocytosis and act as a water channel. Microtubules appear important for the vesicle trafficking. When inactivated, AQPs are returned to the endosomes via protein kinase C activation (Deen and van Os, 1998). Required for this intra-cellular trafficking mechanism are PKA, several PKA-anchoring proteins, and the AVP-induced phosphorylation of AQP2 by PKA at serine 256, which resides on the cytoplasmic carboxyl terminus (Fushimi *et al.*, 1997; Klussmann *et al.*, 1999). Furthermore, vesicle-associated membrane proteins, co-localized with AQP2, facilitate the vesicle trafficking process (King *et al.*, 2000). Wild-type AQP signals are expressed on the apical membrane of oocytes upon stimulation, whereas non-phosphorylated mutant AQPs remain in the cytosolic compartment (Kamsteeg *et al.*, 2000). Recent stoichiometry studies indicate that in principal cells of CDs, three out of four monomers of AQP2 must be phosphorylated prior to initiation of trafficking from the subapical endosome to the apical membrane (Kamsteeg *et al.*, 2000). Although the precise mechanisms by which phosphorylation induces vesicle trafficking are unclear, cytoskeletal structures such as microtubules (transport and the apical polarization of AQP2) and actin microfilaments (vesicle access to the plasma membrane) appear important (Brown *et al.*, 1998). When the AQP2 channel becomes inactive, it returns to the subapical vesicles. In contrast, AQP3 and perhaps AQP4 are constitutively expressed in basolateral membranes and regulate water exit.

In mammalian kidneys, vasopressin regulates AQP2 expression in two stages: first, short-term (minutes) regulation of CD water transport via enhancement of AQP trafficking and second, longer-term (hours) modulation of CD water permeability by up-regulation of AQP2 mRNA and protein (Hayashi *et al.*, 1994; Knepper, 1997; van Os and Deen 1998). In addition, AQP expression appears to be modulated by various physiological and pathological conditions of the kidney (Marples *et al.*, 1999; Knepper, 1998) independent of vasopressin (Nielsen *et al.*, 1999). Hereditary nephrogenic diabetes insipidus (NDI) occurs by mutations of the genes encoding either the vasopressin V_2 receptor or AQP2 (Kamsteeg *et al.*, 1999; Agre, 2000). Acquired NDI, evoked by lithium, bilateral ureteral obstructions, and chronic hypokalaemia, produces a marked reduction in AQP2 expression. Polyuria and down-regulation of AQP2 have been noted in various forms of experimental and acquired NDI, whereas AQP2 expression is elevated in the water retention caused by heart failure or pregnancy (Marples *et al.*, 1999). In general, the urine-concentrating capacity of avian kidneys is lower than that of mammals, particularly rodents (Braun, 1980). It will be interesting to investigate whether AQP2 expression in medullary cones determines the osmotic gradient and countercurrent urine-concentrating capacity among species that live in different habitats.

6.4 *Knockout mice and pathology*

Impairment in AQP genes induces a functional deficiency in water metabolism. Since AQP1 is widely expressed in proximal tubules and DLs, a lack of AQP1 is expected to evoke low Na concentration of tubular urine because of impaired water absorption and subsequent impaired Na homeostasis (Verkman, 1999). Moreover, the loss of water permeability of the DL and vasa recta should lead to insufficient development of the osmotic gradient along the medulla and, hence, impair urine concentration. Surprisingly, the renal phenotype of the AQP1 null family and the appearance and survival of AQP1 knockout mice are grossly normal (Preston *et al.*, 1994; Ashcroft, 2000), suggesting that a para-cellular water pathway independent of the transcellular water channel compensates for the defect. Water-deprived AQP1 knockout mice (Verkman, 1999) or AQP1-deficient humans (King *et al.*, 2000), however, cannot conserve water sufficiently in response to water deprivation; hence, urine osmolality remains low, while plasma osmolality increases (Ma *et al.*, 1998; Verkman, 1999). In human hereditary (familial) NDI, water uptake by the renal tubules and urine-concentrating ability are impaired, and the kidneys excrete a large amount of hypotonic urine. The NDI person exhibits thirst sensation, dehydration, and low plasma volume. Although NDI is most commonly caused by impaired ADH receptors, it has been clarified that mutation of the AQP2 water channel gene (premature termination of translation) evokes autosomal recessive NDI (Deen *et al.*, 1994; Mulders *et al.*, 1998). In *Xenopus* oocytes, mutant AQP2 proteins as encoded in this form of NDI show impairment in routing to the plasma membrane (Mulders *et al.*, 1998). AQP4 knockout mice also show a slight impairment of urine concentration (Verkman, 1999).

6.5 *Aquaporin water channels in non-mammalian species*

More than 23 AQP homologues known as 'green AQPs' have been cloned from plants and may be important for absorbing water via roots. AQP homologues that structurally resemble AQP1 and aquaglyceroporin homologues have also been found in bacteria, yeast and unicellular algae (for review, see Chrispeels and Agre, 1994; Borgnia *et al.*, 1999). In invertebrates, AQPs in insects have been studied. AQP homologues permit transfer of water from the midgut to the Malpighian tubules (Beuron *et al.*, 1995). As to AQPs in non-mammalian vertebrates, although it is anticipated that all vertebrate kidneys possess AQPs, AQP isoforms have been cloned from only a few species of fish, amphibians and birds.

Teleost fish and amphibians. It has been reported that AQP3-homologue mRNA is expressed in the gill, oesophagus and rectal tissues of European eels (Lignot *et al.*, 2001; Cutler and Cramb, 2000). Expression is high in the gill of FW eels, whereas the mRNA levels decrease after the eels are transferred to sea-water (SW). Killifish AQP (recently posted in GenBank; Virkki *et al.*, 2001) homologous to MIP26 has been cloned from the eye lens. AQP water channels and glyceroAQPs have also been cloned from plants, bacteria and insects; but this subject is beyond the scope of this brief review. Although it is anticipated that all vertebrate kidneys possess AQP homologues, AQP isoforms have been cloned in only a limited number of species.

Three AQP homologue cDNAs have been cloned from amphibians: 1) toad bladder AQP (AQP-TB) from the urinary bladder of a toad, *Bufo marinus* (Siner *et al.*, 1996), 2) frog AQP-CHIP (FA-CHIP) from the urinary bladder of a frog, *Rana esculenta* (Abrami *et al.*, 1994), and 3) MIP from frog lens (fMIP) (Austin *et al.*, 1993). The AQP-TB has overall amino acid identity with FA-CHIP (76%), human AQP1 (61%) and rat AQP2 (44%). The homology is higher near the NPA box motifs and lower in the carboxyl terminal domain. Differing from AQP2, however, AQP-TB is expressed abundantly in various tissues, including kidney, skin, lung, skeletal muscles, brain and urinary bladder. Dehydration increases AQP-TB expression in the toad urinary bladder. Expression of AQP-TB protein in *Xenopus laevis* oocytes, however, failed to increase oocyte membrane water permeability (Siner *et al.*, 1996). In contrast, *in vitro* transcribed cRNAs encoding FA-CHIP injected into *Xenopus* oocytes evoked a marked increase in osmotic water permeability (Abrami *et al.*, 1994; Abrami *et al.*, 1995). FA-CHIP has 78% identity with rat and human AQP1, suggesting that the amphibian AQP homologue may be more similar to AQP1. Function of fMIP has not been studied.

Birds. As discussed earlier, the diffusional water permeability of *Coturnix* quail CDs is considerable and is only modestly increased by AVT, whereas adenylate cyclase activity exists abundantly (Nishimura *et al.*, 1996). This may indicate that 1) basal diffusional water transport, perhaps via the AVT-independent water channel, exists in the *Coturnix* CD; and 2) AVT increases water permeability by a cAMP mechanism, via the AVT-dependent water channel. We have molecular and immunochemical evidence that at least two AQP molecules (AQP2 and AQP4) exist in the medullary cone of the avian kidney.

AQP4-homologue channel. Using a cloning method based on reverse transcription-polymerase chain reaction (RT-PCR), we recently cloned two AQPs from medullary cones of water-deprived Japanese quail, *Coturnix coturnix*: 1) an AQP4 homologue (referred to as qAQP4) and 2) a cDNA fragment of AQP2 (qAQP2) (Cui *et al.*, 2001, and unpublished). The total RNA was extracted using a Trizol reagent (Life Technologies; NY, USA) from homogenized tissues. First-strand cDNA was synthesized from total RNA with either random or oligo dT primers, and AQP-related sequences were amplified by the AdvanTag PCR Kit (CLONETECH Lab.; Palo Alto, CA) with a pair of degenerate primers selected from two conserved regions of the MIP family (MIP-F 5′-(T/C)TIAA(T/C)CCIGCIGTIAC-3′ and MIP-R 5′-AAI(G/C)(T/A)IC(G/T)IGCIGG(A/G)T-3′). The PCR products were subcloned into plasmid pDR-TOPO for sequencing. Individual clones having the expected base pair size,

~510 bp (370 + 140 bp linker arms of the vector), were further amplified (T7 primer, M13 reverse primer) and sequenced. The cDNAs for the 5′ and 3′ regions were determined using a commercial kit (Smart RACE, CLONETECH Lab.), and then a full length of cDNA was obtained. The qAQP4 consists of 335 amino acids; and the alignment indicates 77.7% amino acid identity among qAQP4, rat AQP4, human (h) AQP4-L and hAQP4-C2. The qAQP has a longer amino-terminal than rAQP4. Northern blot analysis demonstrated clear mRNA hybridization to qAQP4 cDNA in lung tissue. To determine the water channel function of qAQP4, we examined the expression of qAQP4 cRNA, transcribed *in vitro*, in the stage V and VI oocytes expression system. Oocytes injected with 5 ng of hAQP1 cRNA (control) and incubated in hypo-osmotic Barth's buffer showed marked swelling within 1–2 min, whereas oocyte volume change evoked by qAQP4 was only modest.

AQP2-homologue channel. Using two immunohistochemical methods (conjugation with biotinylated goat anti-mouse IgG and with horseradish peroxidase-labelled streptavidin and FITC-labelled sheep anti-rabbit IgG), we found that cortical and medullary CDs from normally hydrated quail showed positive staining at the apical side of CDs (*Figure 4*). The intensity and area (apical and subapical) of staining increased in the water-deprived group and were most dominant in the water-deprived/AVT-treated quail. Furthermore, the AQP2 antibody recognized proteins of approximately 29 kDa (predicted molecular size) and ~38 kDa (possible glycosylated form) extracted from the medullary cones. The relative densities of the bands at ~29 kDa recognized by AQP2 after water deprivation/AVT treatment suggest that quail kidney CDs possess AQP2-homologue water channels regulated by AVT. AQP signals were found in cultured CD cells that had been manually dissected, dispersed by collagenase, and cultured for a few days. Application of 10^{-5} M of AVT to the medium evoked the translocation of fluorescent signals toward the membrane regions (Zhang *et al.*, unpublished). The sequence of the qAQP2 cDNA fragment was also determined as above, and this fragment contains 90% residues identical to the corresponding region of hAQP2 and 82% residues identical to the rat sequence.

6.6 *Evolutionary perspectives*

During the phylogenetic advancement of vertebrates, the evolution of distal nephrons enabled vertebrates to adapt to hypo-osmotic environments by selectively absorbing ions without the accompaniment of water. The fundamental transport properties of the diluting segment evolved during an early stage of vertebrate evolution, enabling the formation of diluted urine in FW fish and aquatic amphibians. In contrast, the kidney plays a major role in terrestrial life for water conservation. In birds, in the thin DL, epithelial cells show morphological heterogeneity. The upper segment of the DL shows no net volume flux and has low diffusional and osmotic water permeability, whereas its permeability to Na^+ and Cl^- is high and nearly equal, suggesting that NaCl is likely to enter the DL unaccompanied by water. The TAL of the loop of Henle exhibits the characteristics of a diluting segment and provides an energy source for operating countercurrent multiplier mechanism. Hence, in bird kidneys, a countercurrent multiplier system utilizes recirculation of a single solute (NaCl). The single effect may be enhanced by counterflow, physical factors, and the graded hairpin turns of the loop (*Figure 1*). Therefore, although the epithelial

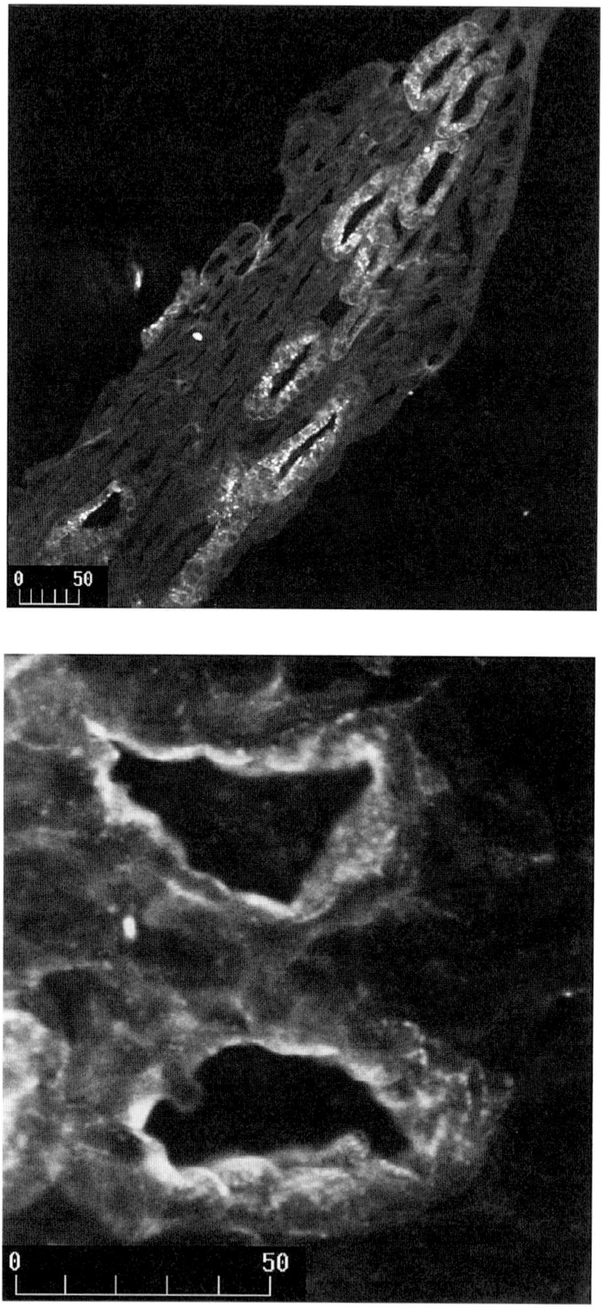

Figure 4. ***Top:*** *A cross section of the medullary cone from a 16-week-old, water-deprived, vasotocin-treated (5.2 mU kg*$^{-1}$*, i.p.)* Coturnix *quail. Medullary cone sections were incubated with anti-rat AQP2 rabbit serum and visualized with FITC-conjugated goat anti-rabbit IgG. Fluorescent signals are seen in CD cells by a laser scanning confocal microscope. Scale bar = 50 μm.* ***Bottom:*** *Higher magnification of the medullary cone section from a 16-week-old, water-deprived, vasotocin-treated (5.2 mU kg*$^{-1}$*, i.p.)* Coturnix *quail. Fluorescent staining is localized at the apical side of CD epithelial cells. Scale bar = 50 μm.*

transport mechanism essential for urine concentration such as the diluting segment evolved at an early phylogenetic stage, kidneys other than from birds and mammals cannot form concentrated urine because of the lack of architectural structures.

A recent study by Liu and coworkers (Liu *et al.*, 2001) indicates that the architectural organization and function of the renal medulla and, hence, the urine-concentrating mechanism of neonatal rats differ from those of mature rats and rather resemble those of avian kidneys. In rat neonatal kidneys, **first**, the thin AL is absent (neither structure nor function), and the entire AL shows the transport characteristics of a TAL expressing a Na-K-2 Cl cotransporter. After birth, AL cells undergo transformation/apoptosis and express a chloride channel specific to the thin AL (Uchida *et al.*, 1995), whereas the expression of a Na-K-2 Cl co-transporter diminishes. **Second,** the DL has low water permeability and does not express AQP1. **Third,** the urea transporter (immunoreactive UT-A1) and urea permeability are absent in the inner medullary CD, indicating that urea is not contributing to the formation of a medullary osmotic gradient. Indeed, human and rat neonates have a lower urine-concentrating capacity than adults (Spitzer and Schwartz, 1992). **Fourth,** although AQP2 is expressed in neonatal rat CDs, vasopressin's effect on diffusional water permeability is only modest. In rats, expressions of AQP1, 3 and 4 plateau before birth, whereas AQP2 mRNA continues to increase after birth (Yamamoto *et al.*, 1997). In newborn rabbits, in which the structure of the medulla is still immature, the medullary osmotic gradient is also low (Horster, 1978). Likewise, the structure and function of the distal nephrons of avian kidneys continue to mature after birth

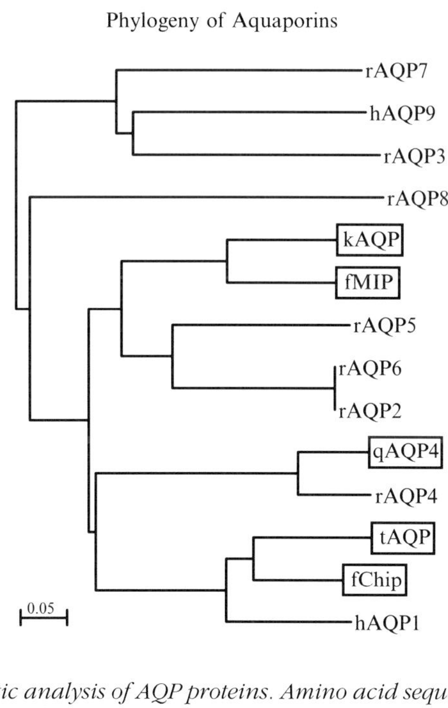

Figure 5. *Phylogenetic analysis of AQP proteins. Amino acid sequences are aligned and analysed by a computerized Clustal X program using a similarity index. Aquaglyceroporins from plants and microorganisms are not included. r, rat; h, human; k, killifish; f, frog; q, quail; t, toad; see text for details.*

(Gabrielli *et al.*, 2000; Narbaitz and Kacew, 1978). Also, during development, water fluxes through the yolk sac, gastrointestinal tract and kidneys (proximal nephrons develop earlier than distal nephrons) should be high via water channels; but AQP expression in developing kidneys is not known. The similarities between maturing mammalian kidneys and avian kidneys may suggest that they share a common ancestral urine-concentrating mechanism. To examine AQP expressions in embryonic avian kidneys will further characterize the role of AQPs in urine-concentrating mechanisms during ontogeny and phylogeny.

A dendrogram for AQP phylogeny calculated with Clustal X based on similarity scores is given in *Figure 5*. Although many AQP isoforms have been reported, the structures of vertebrate AQPs reported by Agre's group and Sasaki's group (see 6.2) are used. AQPs from invertebrates and plants are excluded from this figure. The top part of the figure represents aquaglyceroporins such as AQP7, 9 and 3. It appears that amphibian AQPs are closely related to human or rat AQP1 and that qAQP4 and rAQP4 are more closely related, as expected. Frog MIP (fMIP), from the frog lens, has a structure similar to MIP, but its function has not been examined. Interestingly, AQP4 diverged much earlier than AQP 2. During fetal development, AQP1, AQP3 and AQP4 appear to mature earlier than AQP2 (Yamamoto *et al.,* 1997). The expression of AQP2 mRNA continues to increase after birth. AQP water channels are evolutionarily old and are present in plants, invertebrates and vertebrates. AQP1, AQP3 and AQP4 may be fundamental in renal water handling, whereas AQP2 may have co-evolved with the renal tubule action of vasotocin.

References

Abrami, L., Simon, M., Rousselet, G. and Berthonaud, V. (1994) Sequence and functional expression of an amphibian water channel, FA-CHIP: a new member of the MIP family. *Biochim. Biophys. Acta* **1192**: 147–151.

Abrami, L., Capurro, C., Ibarra, C., Parisi, M., Buhler, J.-M. and Ripoche, P. (1995) Distribution of mRNA encoding the FA-CHIP water channel in amphibian tissues: effects of salt adaptation. *J. Membrane Biol.* **143**: 199–205.

Agre, P. (2000) Aquaporin water channels in kidney. *J. Am. Soc. Nephrol.* **11**: 764–777.

Ashcroft, F.M. (2000). Water channels. In: *Ion Channels and Disease* (ed F.M. Ashcroft). Academic Press, San Diego, pp. 337–345.

Austin, L.R., Rice, S.J., Baldo, G.J., Lange, A.J., Haspel, H.C. and Mathias, R.T. (1993) The cDNA sequence encoding the major intrinsic protein of frog lens. *Gene.* **124**: 303–304.

Bankir, L. and de Rouffignac, C. (1985) Urinary concentrating ability: insights from comparative anatomy. *Am. J. Physiol.* **249**: R643–R666.

Beuron, F., Le Caherec, F., Guillam, M.T., Cavalier, A., Garret, A., Tassan, J.P., Delamarche, C., Schultz, P., Mallouh, V., Rolland, J.P., Hubert, J.F., Gouranton, J. and Thomas, D. (1995) Structural analysis of a MIP family protein from the digestive tract of *Cicadella viridis. J. Biol. Chem.* **270**: 17414–17422.

Borgnia, M., Nielsen, S., Engand, A. and Agre, P. (1999) Cellular and molecular biology of the aquaporin water channels. *Annu. Rev. Biochem.* **68**: 425–458.

Braun, E.J. (1980) Renal osmoregulation. In: *Avian Endocrinology* (eds A. Epple and M.H. Stetson). Academic Press, New York, pp. 499–516.

Braun, E.J. (1993) Renal function in birds. In: *New Insights in Vertebrate Kidney Function* (eds J.A. Brown, R.J. Blament and J.C. Rankin). Cambridge University Press, Cambridge, pp. 167–168.

Braun, E.J. and Dantzler, W.H. (1997) Vertebrate renal system. In: *Handbook of Physiology*, Section 13: Comparative Physiology (ed. W.H. Dantzler). Oxford University Press, New York, Vol. 1, pp. 481–576.

Braun, E.J. and Reimer, P.R. (1988) Structure of avian loop of Henle as related to countercurrent multiplier system. *Am. J. Physio*l. **255** (*Renal Fluid Electrolyte Physiol.* 24): F500–F512.

Brown, D., Katsura, T. and Gustafson, C.E. (1998) Cellular mechanisms of aquaporin trafficking. *Am. J. Physiol.* **275** (*Renal Physiol.* 44): F328–F331.

Burg, M. and Green, N. (1973a) Function of the thick ascending limb of Henle's loop. *Am. J. Physiol.* **224**: 659–668.

Burg, M. and Green, N. (1973b) Effect of ethacrynic acid on the thick ascending limb of Henle's loop. *Kidney Int.* **4**: 301–308.

Casotti, G., Lindberg, K.K. and Braun, E.J. (2000) Functional morphology of the avian medullary cone. *Am J. Physiol.* **279**: R1722–R1730.

Chrispeels, M.J. and Agre, P. (1994) Aquaporins: water channel proteins of plant and animal cells. *TIBS* **19**: 421–425.

Cui, Y., Bolan, J., Wang, W., Zhang, L., Nishimura, H. and Fan, Z. (2001) Cloning of cDNAs encoding avian aquaporin water channels from quail kidney. Biophys. Soc. 45th Ann. Meeting, Boston, MA (Abstr. 2019).

Cutler, C. P. and Cramb, G. (2000) Water transport and aquaporin expression in fish. In: *Molecular Biology and Physiology of Water and Solute Transport* (eds S. Hohmann and S. Nielsen). Academic Press, New York, pp. 433–441.

Dantzler, W.H. and Braun, E.J. (1980) Comparative nephron function in reptiles, birds, and mammals. *Am. J. Physiol.* **239**: R197–R213.

Deen, P.M.T. and van Os, C.H. (1998) Epithelial aquaporins. *Cure. Opin. Cell Biol.* **10**: 435–442.

Deen, P.M.T., Verdijk, M.A.J., Knoers, N.V.A.M., Wieringa, B., Monnens, L.A.H., Os, C.H.V. and Oost, B.A.V. (1994) Requirement of human renal water channel aquaporin-2 for vasopressin-dependent concentration of urine. *Science* **264**: 92–95.

de Rouffignac, C., Roinel, N. and Elalouf, J.M. (1993) Comparative effects of peptide hormones on water and electrolyte transport along the proximal and distal tubules of the mammalian nephron. In: *New Insights in Vertebrate Kidney Function* (eds J.A. Brown, R.J. Balment and J.C. Rankin). Society for Exprerimental Biology Seminar Series 52. Cambridge University Press, Cambridge, pp. 281–309.

Dietl, P. and Stanton, B.A. (1993) The amphibian distal nephron. In: *New Insights in Vertebrate Kidney Function* (eds J.A. Brown, R.J. Balment, and C.J. Rankin). Society for Exprerimental Biology Seminar Series 52. Cambridge University Press, Cambridge, pp. 115–134.

Echevarria, M., Windhager, E.E., Tate, S.S. and Frindt, G. (1994) Cloning and expression of AQP3, a water channel from the medullary collecting duct of rat kidney. *Proc. Natl Acad. Sci. USA* **91**: 10997–11001.

Friedman, P.A. and Hebert, S.C. (1990) Diluting segment in kidney of dogfish shark, I. Localization and characterization of chloride absorption. *Am. J. Physiol.* **258**: R398–R408.

Fushimi, K., Uchida, S., Hara, Y., Hirata, Y., Marumo, F. and Sasaki, S. (1993) Cloning and expression of apical membrane water channel of rat kidney collecting tubule. *Letters to Nature* **361**: 549–552.

Fushimi, K., Sasaki, S., Yamamoto, T., Hayashi, M., Furukawa, T., Uchida, S., Kuwahara, M., Ishibashi, K., Kawasaki, M., Kihara, I. and Marumo, F. (1994) Functional characterization and cell immunolocalization of AQP-CD water channel in kidney collecting duct. *Am. J. Physiol.* **267** (*Renal Fluid Electrolyte Physiol.* 36): F573–F582.

Fushimi, K., Sasaki, S. and Marumo, F. (1997) Phosphorylation of serine 256 is required for cAMP-dependent regulatory exocytosis of the aquaporin-2 water channel. *J. Biol. Chem.* **272**: 14800–14804.

Gabrielli, M.G., Materazzi, G. and Menghi, G. (2000) The metanephros of the quail embryo. Developmental expression of carbonic anhydrase investigated by multiple approaches. *J. Anat.* **196**: 31–40.

Goldstein, D.L. and Braun, E.J. (1989) Structure and concentrating ability in the avian kidney. *Am. J. Physiol.* **256** (*Regulatory Integrative Comp. Physiol.* 25): R501–R509.

Goldstein, D.L., Reddy, V. and Plaga, K. (1999) Second messenger production in avian medullary nephron segments in response to peptide hormones. *Am. J. Physiol.* **276** (*Regulatory Integrative Comp. Physiol.* 45): R847–R854.

Hall, D.A. and Varney, D.M. (1980) Effect of vasopressin on electrical potential difference and chloride transport in mouse medullar thick ascending limb of Henle's loop. *J. Clin. Invest.* **66**: 792–802.

Han, Z., Wax, M.B. and Patil, R.V. (1998) Regulation of aquaporin-4 water channels by phorbol ester-dependent protein phosphorylation. *J. Biol. Chem.* **273**: 6001–6004.

Hayashi, M., Sasaki, S., Tsuganezawa, H., Monkawa, T., Kitajima, W., Konishi, K., Fushimi, K., Marumo, F. and Saruta, T. (1994) Expression and distribution of aquaporin of collecting duct are regulated by vasopressin V_2 receptor in rat kidney. *J. Clin. Invest.* **94**: 1778–1783.

Hebert, S.C. and Friedman, P.A. (1990) Diluting segment in kidney of dogfish shark. II. Electrophysiology of apical membranes and cellular resistances. *Am. J. Physiol.* **258**: R409–R417.

Hebert, S.C., Culpepper, R.M. and Andreoli, T.E. (1981) NaCl transport in mouse medullary thick ascending limbs. II. ADH enhancement of transcellular NaCl cotransport; origin of transepithelial voltage. *Am. J. Physiol.* **241**: F432–F442.

Horster, M. (1978) Loop of Henle functional differentiation in vitro perfusion of the isolated thick ascending segment. *Pflugers Arch.* **378**: 15–24.

Imai, M., Hayashi, M. and Araki, M. (1984) Functional heterogeneity of the descending limbs of Henle's loop. I. Internephron heterogeneity in the hamster kidney. *Pflugers Arch.* **402**: 385–399.

Imai, M., Taniguchi, J. and Yoshitomi, K. (1988) Transition of permeability properties along the descending limb of long-loop nephron. *Am. J. Physiol.* **254**: F323–F328.

Ishibashi, K., Kuwahara, M., Gu, Y., Kageyama, Y., Tohsaka, A., Suzuki, F., Marumo, F. and Sasaki, S. (1997) Cloning and functional expression of a new water channel abundantly expressed in the testis permeable to water, glycerol, and urea. *J. Biol. Chem.* **272**: 20782–20786.

Johnson, O.W. and Mugaas, J.N. (1970) Some histological features of avian kidneys. *Am. J. Anat.* **127**: 423–436.

Jung, J.S., Preston, G.M., Smith, B.L., Guggino, W.B. and Agre, P. (1994) Molecular structure of the water channel through aquaporin CHIP. *J. Biol. Chem.* **269**: 14648–14654.

Kamsteeg, E.J., Wormhoudt, T.A.M., Rijss, J.P.L., van Os, C.H. and Deen, P.M.T. (1999) An impaired routing of wild-type aquaporin-2 after tetramerization with an aquaporin-2 mutant explains dominant nephrogenic diabetes insipidus. *EMBO J.* **18**: 2394–2400.

Kamsteeg, E.J., Heijnen, I., van Os, C.H. and Deen, P.M.T. (2000) The sub-cellular localization of an aquaporin-2 tetramer depends on the stoichiometry of phosphorylated and nonphosphorylated monomers. *J. Cell Biol.* **151**: 919–930.

King, A.S. and King, D.Z. (1979) Avian morphology: general principles. In: *Form and Function in Birds* (eds A.S. King and J. McLelland). Academic Press, New York. Vol. 1, pp. 1–38.

King, L.S.M., Yasui, M. and Agre, P. (2000) Aquaporins in health and disease. *Molecular Medicine Today* **6**: 60–65.

Klussmann, E., Maric, K., Wiesner, B., Beyermann, M. and Rosenthal, W. (1999) Protein kinase A anchoring proteins are required for vasopressin-mediated translocation of aquaporin-2 into cell membranes of renal principal cells. *J Biol. Chem.* **274**: 4934–4938.

Knepper, M.A. (1997) Molecular physiology of urinary concentrating mechanism: regulation of aquaporin water channels by vasopressin. *Am. J. Physiol.* **272** (*Renal Physiol.* 41): F3–F12.

Knepper, M.A. (1998) Long-term regulation of urinary concentrating capacity. *Am. J. Physiol.* **275** (*Renal Physiol.* 44): F332–F333.

Kondo, Y. and Imai, M. (1987) Effects of glutaraldehyde fixation on renal tubular function. I. Preservation of vasopressin-stimulated water and urea pathways in rat papillary collecting duct. *Pflugers Arch.* **408**: 479–483.

Kriz, W., Schiller, A., Kaissling, B. and Taugner, R. (1980) Comparative and functional aspects of thin loop limb ultrastructure. In: *Functional Ultrastructure of the Kidney*. Academic Press, London, pp. 39–250.

Laycock, J.F. and Hanoune, J. (1998) From vasopressin receptor to water channel: intracellular traffic, constraint and by-pass. *J. Endocrinol.* **159**: 361–372.

Lignot, J.H., Cutler, C.P., Hazon, N. and Gramb, G. (2001) Immunolocalization of aquaporin 3 in the gill and the gastrointestinal tract of the European eel *Anguilla anguilla*. In: Symposium 'Drink or Die' (org N. Hazon and G. Flik). Sorrento, Italy.

Liu, W., Morimoto,T., Kondo, Y., Iinuma, K., Uchida, S. and Imai, M. (2001) 'Avian-type' renal medullary tubule organization causes immaturity of urine-concentrating ability in neonates. *Kidney Int.* **60**: 680–693.

Logan, A.G., Moriarty, R.J. and Rankin, J.C. (1980) A micropuncture study of kidney function in the river lamprey, *Lampetra fluviatilis*, adapted to fresh water. *J. Exp. Biol.* **85**: 137–147.

Ma, T., Yang, B., Gillespie, A., Carlson, E.J., Epstein, C.J. and Verkman, A.S. (1998) Severely impaired urinary concentrating ability in transgenic mice lacking aquaporin-1 water channels. *J. Biol. Chem.* **273**: 4296–4299.

Marples, D., Frøkiaer, J. and Nielsen, S. (1999) Long-term regulation of aquaporins in the kidney. *Am. J. Physiol.* **276**: F331–F339.

Mitra, A. K. (2001) Three-dimensional organization of the aquaporin water channel: what can structure tell us about function? *Vitam. Horm.* **62**: 133–166.

Miwa, T. and Nishimura, H. (1986) Diluting segment in avian kidney. II. Water and chloride transport. *Am. J. Physiol.* **250** (*Regulatory Integrative Comp. Physiol.* 19): R341–R347.

Mulders, S.M., Bichet, D.G., Rijss, J.P.L., Kamsteeg, E., Arthus, M., Lonergan, M., Fujiwara, M., Morgan, K., Leijendekker, R., Sluijs, P.V.D., van Os, C.H. and Deen, P.M.T. (1998) An aquaporin-2 water channel mutant which causes autosomal dominant nephrogenic diabetes insipidus is retained in the golgi complex. *J. Clin. Invest.* **102**: 57–66.

Narbaitz, R. and Kacew, S. (1978) Ultrastructural and biochemical observations on the metanephros of normal and cultured chick embryos. *Anat. Embryol.* **155**: 95–105.

Nicholson, J.K. and Kendall, M.D. (1983) The fine structure of dark or intercalated cells from the distal and collecting tubules of avian kidneys. *J. Anat.* **136**: 145–156.

Nielsen, S., Pallone, T. and Smith, B.L. (1995) Aquaporin-1 water channels in short and long loop descending thin limbs and in descending vasa recta in rat kidney. *Am. J. Physiol.* **268**: F1023–F1037.

Nielsen, S., Kwon, T., Christensen, B.M., Promeneur, D., Frøkiaer, J. and Marples, D. (1999) Physiology and pathophysiology of renal aquaporins. *J. Am. Soc. Nephrol.* **10**: 647–663.

Nishimura, H. (1989) Urine concentration in birds. In: *Progress in Avian Osmoregulation* (eds M.R. Hughes and A. Chadwick). Scientific Session of the Leeds Physiological and Literary Society, London, pp. 225–242.

Nishimura, H. (1993) Countercurrent urine concentration in birds. In: *New Insight in Vertebrate Kidney Function* (eds J.A. Brown, R.J. Balment and J.C. Rankin). Cambridge University Press, Society for Experimental Biology Seminar Series 52, pp. 189–212.

Nishimura, H., Imai, M. and Ogawa, M. (1983) Sodium chloride and water transport in the renal distal tubule of the rainbow trout. *Am. J. Physiol.* **244**: F247–F254.

Nishimura, H., Imai, M. and Ogawa, M. (1986) Diluting segment in avian kidney, I. Characterization of transepithelial voltages. *Am. J. Physiol.* **242**: F331–F339.

Nishimura, H., Koseki, C., Imai, M. and Braun, E.J. (1989) Sodium chloride and water transport in the thin descending limb of Henle of the quail. *Am. J. Physiol.* **257** (*Renal Fluid Electrolyte Physiol.* 26): F994–F1002.

Nishimura, H., Koseki, C. and Patel, T.B. (1996) Water transport in collecting ducts of Japanese quail. *Am. J. Physiol.* **271** (*Regulatory Integrative Comp. Physiol.* 40): R1535–R1543.

Nishimura, H., Xu, W., Fan, Z. and Sasaki, S. (2000) Aquaporin-homologue water channel in avian kidney. Experimental Biology annual meeting. LB28, San Diego, pp. 5.

Nishimura, H., Fan, Z. and Zhang, L. (2001) Renal handling of water in vertebrates. In: *Perspective in Comparative Endocrinology* (eds H. J.T. Goos, R.K. Rastogi, H. Vaudry and R. Pierantoni). Monduzzi editore, International proceedings division, pp. 733–743.

Oberleithner, H., Guggino, W. and Giebisch, G. (1982) Mechanism of distal tubular chloride transport in *Amphiuma* kidney. *Am. J. Physiol.* **242**: F331–F339.

Osono, E. and Nishimura, H. (1994) Control of sodium and chloride transport in the thick ascending limb in the avian nephron. *Am. J. Physiol.* **267** (*Regulatory Integrative Comp. Physiol.* 36): R455–R462.

Preston, G.M. and Agre, P. (1991) Isolation of the cDNA for erythrocyte integral membrane protein of 28 kilodaltons: member of an ancient channel family. *Proc. Natl Acad. Sci. USA* **88**: 11110–11114.

Preston, G.M., Smith, B.L., Zeidel, M.L., Moulds, J.J. and Agre, P. (1994) Mutations in aquaporin-1 in phenotypically normal humans without functional CHIP water channels. *Science* **265**: 1585–1587.

Rocha, A.S. and Kokko, J.P. (1973) Sodium chloride and water transport in the medullary thick ascending limb of Henle. *J. Clin. Invest.* **52**: 612–623.

Siner, J., Páredes, A., Hosselet, C., Hammond, T., Strange, K. and Harris, H.W. (1996) Cloning of an aquaporin homologue present in water channel containing endosomes of toad urinary bladder. *Am. J. Physiol.* **270** (*Cell Physiol.* 39): C372–C381.

Spitzer, A. and Schwartz, G. (1992) The kidney during development. In: *Handbook of Physiology: Renal Physiology* (ed. E. Windhager). Oxford University Press, New York, Vol. 1, pp. 475–544.

Stoner, L.C. (1977) Isolated, perfused amphibian renal tubules: the diluting segment. *Am. J. Physiol.* **233**: F438–F444.

Terris, J., Ecelbarger, C.A., Marples, D., Knepper, M.A. and Nielsen, S. (1995) Distribution of aquaporin-4 water channel expression within rat kidney. *Am. J. Physiol.* **269** (*Renal Fluid Electrolyte Physiol.* 38): F775–F785.

Trinh-Trang-Tan, M.M. and Bankir, L. (1998) Integrated function of urea transporters in the mammalian kidney. *Exp. Nephrol.* **6**: 471–479.

Trinh-Trang-Tan, M.M. and Bankir, L. (2000) Renal urea transporters. Direct and indirect regulation by vasopressin. *Exp. Physiol.* **85S**: 243S–252S.

Uchida, S., Sasaki, S., Furukawa, T., Hiraoka, M., Imai, T., Hirata, Y., and Marumo, F. (1993) Molecular cloning of a chloride channel that is regulated by dehydration and expressed predominantly in kidney medulla. *J. Biol. Chem.* **268**: 3821–3824.

Uchida, S., Sasaki, S., Nitta, K., Uchida, K., Horita, S., Nihei, H. and Marumo, F. (1995) Localization and functional characterization of rat kidney-specific chloride channel, CIC-K1. *J. Clin. Invest.* **95**: 104–113.

van Os, C.H. and Deen, P.M.T. (1998) Role of aquaporins in renal water handling: physiology and pathophysiology. *Nephrol. Dial Transplant* **13**: 1645–1651.

Verkman, A.S. (1999) Lessons on renal physiology from transgenic mice lacking aquaporin water channels. *J. Am. Soc. Nephrol.* **10**: 1126–1135.

Verkman, A.S. and Mitra, A.K. (2000) Structure and function of aquaporin water channels, *Am. J. Physiol.* **278**: F13–F28.

Virkki, L.V., Cooper, G.J. and Boron, W.F. (2001) Cloning and functional expression of a MIP (AQP0) homolog from killifish (*Fundulus heteroclitus*) lens. *Am. J. Physiol. (Regulatory, Integrative and Comparative)* **281:** R1994–R2003.

Yamamoto, T. and Sasaki, S. (1998) Aquaporins in the kidney: emerging new aspects. *Kidney Internat.* **54**: 1041–1051.

Yamamato, T., Sasaki, S., Fushimi, K., Ishibashi, K., Yaoita, E., Kawasaki, K., Fujinaka, H., Marumo, F. and Kihara, I. (1997) Expression of AQP family in rat kidneys during development and maturation. *Am. J. Physiol.* **272** (*Renal Physiol.* 41): F198–F204.

Yasui, M., Marples, D., Belusa, R., Eklöf, A.-C., Celsi, G., Nielsen, S. and Aperia, A. (1996) Development of urinary concentrating capacity: role of aquaporin-2. *Am. J. Physiol.* **271** (*Renal Fluid Electrolyte Physiol. 40*): F461–F468.

Yasui, M., Hazama, A., Kwon, T.H., Nielsens, S., Guggino, W.B. and Agre, P. (1999a) Rapid gating and anion permeability of an intracellular aquaporin. *Nature* **402**: 184–187.

Yasui, M., Kwon, T., Knepper, M.A., Nielsen, S. and Agre, P. (1999b) Aquaporin-6: an intracellular vesicle water channel protein in renal epithelia. *Proc. Natl. Acad. Sci. USA* **96**: 5808–5813.

Young, J.Z. (1981) The origin and evolution in birds. In: *The Life of Vertebrates*, 3rd edn. Clarendon Press, Oxford, pp. 386–401.

Zhang, L., Nishimura, H., Cui, Y.J., Fan, Z. and Sasaki, S. (2001) Medullary collecting ducts of coturnix kidney express both vasotocin-dependent and vasotocin-independent aquaporins. *FASEB J.* **15**: part 1, A410.

10

Control of water balance in mammals

R.J. Balment

1. Introduction

Survival of mammals in the desiccating terrestrial environment is very much dependent upon securing stability of body fluid volume and composition. This selective pressure has been associated with the development of sophisticated water conservation measures in the kidney together with elaborate behavioural adaptations including thirst. These adaptive functions are in turn regulated by a complex interplay of hormonal and neuronal control systems sensitive to subtle changes in fluid status.

In mammals, water is broadly 70% of lean body mass, two thirds of which is intracellular while one third is extra-cellular (vascular or interstitial). Although only 25% of extra-cellular fluid is contained in plasma, maintenance of plasma volume is vital for survival with significant loss leading to circulatory collapse and death. Control of water balance in mammals is essentially directed toward securing both plasma volume and intra-cellular fluid volume. In man the turnover of water is usually about 2–3 l per day. The losses of water through respiratory exchange, sweating, gastrointestinal tract and urination are largely balanced by the ingestion of fluid. Compensatory drinking to secure balance is sensitively regulated but it is the evolution of the mammalian metanephric kidney, permitting the production of urine highly hypertonic to plasma, which affords the important capacity for water and solute conservation. The kidney, through its potential to conserve water to a greater extent than all other vertebrates, is the organ that has given terrestrial mammals their freedom of movement away from stable water sources. This chapter will examine aspects of the control of water balance in mammals and the integration of renal and central thirst mechanisms.

2. Renal mechanisms supporting water balance

In contrast to the mesonephric kidneys of fish and amphibians the mammalian metanephric kidney exhibits autoregulation of glomerular filtration rate (GFR).

Osmoregulation and Drinking in Vertebrates, edited by N. Hazon & G. Flik.
© 2002 BIOS Scientific Publishers Ltd, Oxford

Consequently, substantive changes in systemic blood pressure do not greatly perturb GFR, which is maintained at a relatively high, constant rate. The large volume of filtered fluid is largely (65%) reabsorbed in the early proximal tubule, while a renal medullary concentration gradient is established by the selective absorption of solute in the intermediate diluting segment, the loop of Henle. The dilute (hypotonic to plasma) nascent urine emerging from the loop is further modified by regulated selective reabsorption of solute and water in the remaining distal nephron segments. The selective reabsorption of water is largely in response to the action of arginine vasopressin (AVP) the mammalian neurohypophysial antidiuretic hormone (ADH), which in this group replaces the arginine vasotocin (AVT) found in all other non-mammalian vertebrates (Acher, 1996). In mammals the major action of ADH is tubular, promoting increased selective water permeability and consequent water absorption in the distal nephron, particularly the collecting duct. In contrast, the major action of ADH in lower vertebrates is to reduce urine flow through a fall in GFR. Accordingly, we can ascribe a shift in location (distal nephron vs juxtaglomerular (JG) vasculature) and ADH receptor type (V_2 epithelial vs V_1 vascular) associated with the development of the mammalian kidney (Pang *et al.*, 1983). The coincident emergence of the loop of Henle in mammals has also permitted the development of a large standing renal medullary osmotic gradient (largely NaCl and urea), which enhances the ability for distal tubular water reabsorption when ADH acts to increase its water permeability. This gradient may be further enhanced by another action of ADH. The AVP V_2 receptor is present in the thick ascending limb of the loop of Henle (Burgess and Balment, 1993), and activation leads to increased activity of the NaK_2Cl transporter (de Rouffignac *et al.*, 1991).

The importance of the distal tubular action of ADH in mammals and evidence of the extremely low resting water permeability of this segment, is exemplified by the high urine production rate which accompanies absence of ADH. As *Table 1* indicates, the genetically vasopressin-deficient Brattleboro rat, by comparison with its control Long Evans rat, exhibits very high daily urine production rates which approach 75% of body weight. This urine is very dilute at approximately half the concentration of blood plasma (ca. 300 mOsmol kg^{-1}) and contrasts with the urine of vasopressin-replete Long Evans animals, which is three-fold more concentrated than plasma (see *Table 1*). These huge renal losses of water in the Brattleboro rat are compensated by high drinking water intakes. It is also evident for these age-matched male rats that the growth of the Brattleboro rat is greatly reduced by com-

Table 1. *Water balance in the presence and absence of antidiuretic hormone*

	Body weight (g)	Urine volume (ml)	Urine osmolality (mOsm kg^{-1})	Water intake (ml)
Long Evans rat (*n* = 6)	425 ± 11	19.7 ± 0.5	1382 ± 34	48.1 ± 1.7
Brattleboro rat (ADH-deficient) (*n* = 5)	245 ± 18	181.9 ± 13.6	159 ± 10	199 ± 14.4

Values are mean ± SE for 24 h metabolism cage measures.

parison with the wild-type animal, in part reflecting the metabolic cost of this high water turnover.

The water-conserving role of ADH is effected by the insertion of water channels (specifically aquaporin2 – AQP2) into the luminal membrane of distal nephron principal cells (PCs) (Nielson and Agre, 1995). The binding of ADH to its V_2 receptor on the basolateral side of the tubule activates adenylate cyclase and induces a rise in intra-cellular cyclic AMP. This in turn, as illustrated in *Figure 1*, leads to a cascade of intra-cellular events which promotes movement of intramembranous particles (IMPs), the AQP2 channels, into the apical membrane.

These water channels are rapidly recycled back into the cell and will not be re-inserted unless further ADH activation of the intra-cellular cascade occurs. Accordingly, adjustments in secretion and circulating ADH concentration reaching the kidney can effect rapid and very sensitive changes in tubule water permeability and urine volume. Through these tubular actions of ADH, a balance of renal excretion and water availability is achieved without the need to adjust GFR. This contrasts with the position in lower vertebrate groups but is important for the warm-blooded mammals, with their high metabolic rates and heavy reliance upon renal excretion of metabolic waste products.

Other significant hormonal regulators of renal water excretion include the renin–angiotensin system (RAS), adrenal steroids and atrial natriuretic peptide (ANP). Renin release from afferent renal arterioles associated with the JG apparatus, is sensitive to changes in blood volume and general hydrational state (see section 10.4). Intra-renally formed angiotensin II (ANG II) can directly influence proximal tubule solute and fluid absorption (Hall *et al.*, 1990) and thus enhance any further ADH stimulated water reabsorption in the distal nephron. The stimulation of aldosterone secretion

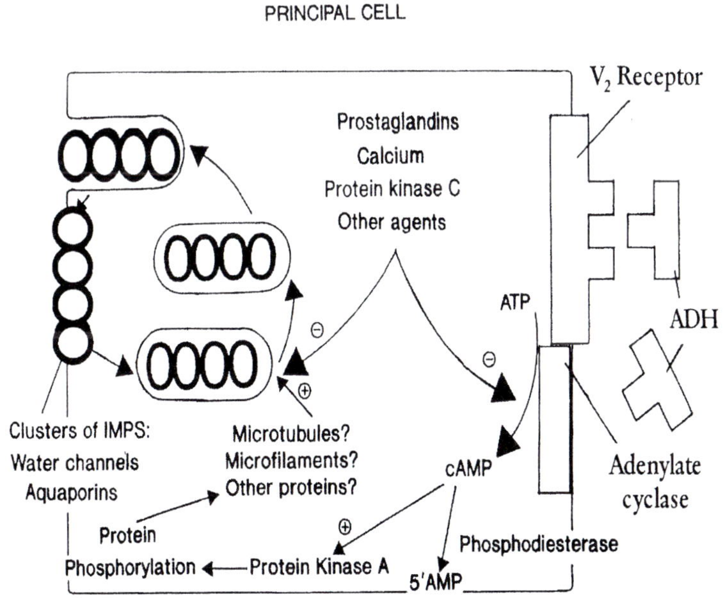

Figure 1. *Model of cellular events following ADH stimulation of the V_2 receptor, resulting in increased collecting duct apical membrane water permeability.*

from the adrenal cortex by ANG II may lead indirectly to further water reabsorption in association with aldosterone's action to promote Na^+ reabsorption in the distal tubule. Adrenal steroids (including cortisol and corticosterone) may also be important in maintaining normal GFR and the water permeability of the distal renal tubule. A characteristic feature of adrenal insufficiency is the reduced ability to excrete a water load. Indeed, following removal of the adrenals in Brattleboro rats (Balment *et al.*, 1977) there is a considerable reduction in urine production (73.9 ± 5.1 to 43.6 ± 4.3 ml 100 g^{-1} body weight (BW) per 24 h, n = 6) and consequent fall in compensatory water intake (79.4 to 47.7 ml 100 g^{-1} BW per 24 h). Adrenal steroids also appear to be essential for the normal action of ADH on the distal nephron.

The response to ADH (AVP) of isolated inner medullary collecting ducts (IMCD) prepared from intact and adrenalectomized rats is shown in *Figure 2*. The ADH response is presented as fmol cAMP generated during 4 min incubation and is expressed per mm tubule length (Wargent and Balment, 1997). In the absence of adrenal steroids the collecting duct segments appear unresponsive and do not show the normal dose-dependent AVP-stimulated increase in cAMP production. This would clearly impair the normal increase in water permeability induced by AVP and thus reduce the ability to conserve water during dehydration.

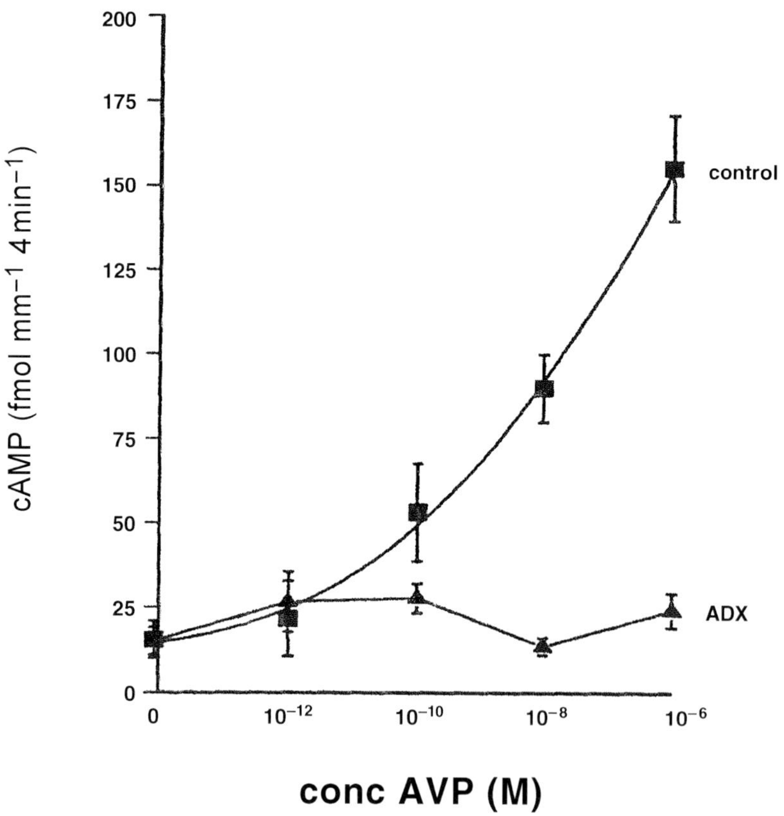

Figure 2. *Effect of adrenalectomy on AVP-induced cAMP accumulation in IMCD segments. Values are means ± SE for 8–10 tubules at each AVP concentration obtained from 3–4 intact (control) or 4-day adrenalectomized rats.*

This interaction between hormone systems to modulate the effectiveness of each other's action is a recurring phenomenon in the mechanisms responsible for maintaining water balance. That is to say, there is a reserve of multiple regulatory inputs to control renal excretion. Accordingly, appropriate renal excretion of water will reflect the balanced regulatory input of a number of systems (see section 10.5). Such apparent redundancy is clearly selected to support a vital physiological function for survival. Accordingly, in recent years it has become apparent that the cardiac hormones, such as ANP, act to promote renal loss of solute and water in hyper-volaemic situations (Ruskoaho, 1992). While these natriuretic peptide actions in part reflect a direct effect on the distal nephron to inhibit sodium and associated water absorption, ANP also reduces RAS activity, inhibits the vasoconstrictor actions of ANG II and ANG II stimulation of adrenal aldosterone secretion.

There are three major members of this natriuretic peptide family in mammals, ANP, brain natriuretic peptide (BNP) and C-type natriuretic peptide (CNP) and these are considered to act through two major types of bioactive receptor subtypes ANP-A and ANP-B (Imura *et al.*, 1992). BNP, though originally isolated from porcine brain, is mainly produced in the heart, while CNP is a neuropeptide. The peptides and their receptors are widely distributed in brain regions involved in drinking behaviour, body fluid homeostasis and blood pressure regulation (Gutkowska *et al.*, 1997). A further potential member of the natriuretic peptide family, urodilatin has also been identified in the kidney (Forssmann, 1995). This potentially affords local paracrine control of renal sodium and water excretion, perhaps under cephalic control (Emmeluth *et al.*, 1992) sensitive to changes in brain circulation sodium concentration.

3. Thirst and compensatory drinking

Our understanding of mammalian thirst mechanisms has been driven by early attempts to assess the separate and relative roles of deviations in the cellular and extra-cellular fluid compartments.

3.1 *Cellular dehydration as a thirst stimulus*

Gilman in 1937 showed that a critical stimulus for drinking is loss of water from inside cells. Administration of hypertonic solutions of solutes to which cell membranes were impermeable stimulated drinking in association with cell shrinkage, as a result of osmotic loss of intra-cellular water. By contrast, administration of hypertonic solutions such as glucose, which can readily enter cells and does not lead to cellular dehydration, stimulated little or no drinking. It thus appears that it is cell shrinkage/dehydration rather than a rise in fluid osmolality per se, which triggers thirst behaviour and drinking. However, the work of Andersson (1978) and others has drawn attention to the possibility that ruminants like goat and sheep, which have a large sodium turnover compared with dog and rat, can detect changes in cerebral spinal fluid (CSF) sodium concentration during dehydration (McKinley, 1991). Of course in practice the largest osmotic component of extra-cellular fluid is sodium and its attendant anions (90–95%), the concentration of which will rise during dehydration, leading to both cellular dehydration and a rise in CSF sodium concentration.

A number of specific osmoreceptor locations for thirst have been proposed, including stomach, gut, hepatic-portal circulation and the central nervous system. The greatest evidence exists for the latter with relatively small volumes (0.5–2 μl) of mildly hypertonic saline (0.8–0.3 M NaCl) injected into the lateral preoptic area eliciting drinking in rat and rabbit. Such studies have indicated that osmoreceptors appear to be widely distributed in the lateral preoptic and lateral hypothalamic areas. The thirst-inducing action of such injections does not always coincide with stimulation of ADH secretion, suggesting that at least part of the osmoreceptors for thirst and ADH secretion are distinct (Rolls and Rolls, 1982). Although the osmoreceptors of the central nervous system respond directly to changes in tonicity of cerebral blood, osmotically sensitive receptors present in the stomach, gut and hepatic-portal blood system, may also feed information to the brain (Rolls and Rolls, 1982).

3.2 *Extra-cellular fluid volume as a thirst stimulus*

Although the amount of fluid in the extra-cellular compartment is much less than cellular fluid volume, this is a vital compartment. Any significant compromise of vascular fluid volume and pressure will lead to reduced performance and potentially circulatory collapse. The exchange of water between the body and external environment is effected through the extra-cellular fluid. The two phases of the extra-cellular compartment are usually in equilibrium, the steady state being maintained by exchange of water and ions across capillary walls. Most control mechanisms for maintaining fluid balance rest inside the vascular space. Stretch receptors and pressure receptors (baroreceptors) are present within the low pressure venous and high pressure arterial system. A decrease in plasma volume detected in this way rapidly stimulates ADH secretion and renin release and consequent renal conservation of water. However, replenishment of the lost fluid can only be achieved through increased drinking. Haemorrhage, that is isotonic fluid loss, selectively depleting the extra-cellular fluid compartment without osmotic effects, leads to increased drinking. A marked appetite for sodium also develops and subsequent ingestion of sodium restores the extra-cellular volume and composition to normal (Stricker, 1981).

Drinking which follows extra-cellular fluid depletion may be triggered by stretch or pressure receptors in the vasculature but also involves the renal RAS. Fitzsimons' classical studies showed that specific reduction in renal perfusion pressure elicited drinking and that renin release was involved (see Fitzsimons, 1969). Although ANG II given systemically can elicit drinking, only small quantities are required if given centrally, where ANG II receptors are found in the circumventricular organs. In this region the blood brain barrier is not complete and potentially receptors can be activated by both peripheral ANG II or centrally generated ANG II. It is now apparent that all components of the RAS are present within the brain and that this may be involved in the central mechanisms modulating thirst and drinking behaviour and ADH secretion (Fitzsimons, 1998).

After substantial blood loss in rat, ANG II levels in the hypothalamus rise independently of increase in plasma ANG II, which could imply a role for brain ANG II in drinking and blood pressure regulation in hypo-volaemia (Phillips *et al.*, 1996). Whether angiotensin peptides derived from brain should be regarded as neurotransmitters or as also performing a paracrine role remains to be established.

Nonetheless, most experimental evidence supports the concept that brain ANG II is important in regulation of blood pressure, sodium homeostasis and blood volume. These are related physiological functions that have direct bearing on maintenance of water balance. Overactivity of the brain RAS has been implicated in the development of hypertension in the spontaneously hypertensive rat (Bannermann *et al.*, 1993). Central ANG II may also play a part in thirst and drinking in response to cellular dehydration (Andersson *et al.*, 1972). Central ANG II in addition to peripherally generated ANG II can also stimulate vasopressin secretion (Saaredra, 1992). Vasopressin in turn may further affect water intake by increasing the sensitivity of neuronal thirst systems (Fitzsimons, 1980). Vasopressin is co-localized with ANG II in the magnocellular neurons of the supraoptic and paraventricular nuclei (SON and PVN), though whether there is co-release and interaction of the peptides on a single target tissue remains unknown. However, it is evident that lack of vasopressin does not affect ANG II-stimulated drinking in the Brattleboro rat.

Large numbers of mammalian species have now been shown to show profound dipsogenic responses to ANG II, as observed in all other vertebrate groups (Takei, 2000) with the possible exception of amphibians, in which percutaneous water absorption may be affected (Hillyard *et al.*, 1998). A rise in endogenous ANG II occurs alongside fluid depletion and ANG II-induced drinking will add to drinking stimulated be cellular dehydration in the rat (Fitzsimons and Simons, 1969). The very high compensatory drinking rates observed in Brattleboro rats are associated with greatly elevated plasma renin activity (Oliver *et al.*, 1982). The interaction of thirst stimuli may vary from species to species. In the dog, for example, a subthreshold dose of ANG II will lower the thirst threshold to hypertonic saline, while in the goat combined ANG II and hypertonic saline elicit greater drinking rates than the added effects when given separately, and in overhydrated rats ANG II is relatively ineffective in stimulating drinking (Rolls and McFarland, 1973). Pharmacological blockade of the RAS does not prevent compensatory drinking in dehydration, which implies, as for the kidney, we have a multiple backup system to control adaptive drinking and ensure survival.

In summary, water deprivation leads to activation of both cellular dehydration and vascular hypo-volaemic pathways to promote thirst and drinking. This is summarized in *Figure 3* which also indicates that the hypo-volaemic component may involve both RAS and RAS-independent contributions.

3.3 *Cessation or termination of drinking*

During water deprivation in the rat and dog both cellular and extra-cellular fluid compartments are depleted. On the basis of deficit replacement studies (Rolls and Rolls, 1982) cellular fluid depletion accounts for 60–70% of the drinking and the extra-cellular fluid depletion 20–30%. Clearly there is danger if there is over expansion of the extra-cellular compartment during compensatory drinking, which could lead to overdilution and possibly water intoxication. In most species systemic changes in osmolality and volume do not occur rapidly enough upon drinking to provide a mechanism for cessation of drinking. However, overhydration does not occur when access to water is restored to dehydrated animals (Adolf, 1950). Oropharyngeal sensors, gastric distension and intestinal or hepatic-portal factors may all contribute to the normal termination of drinking. The operation of these

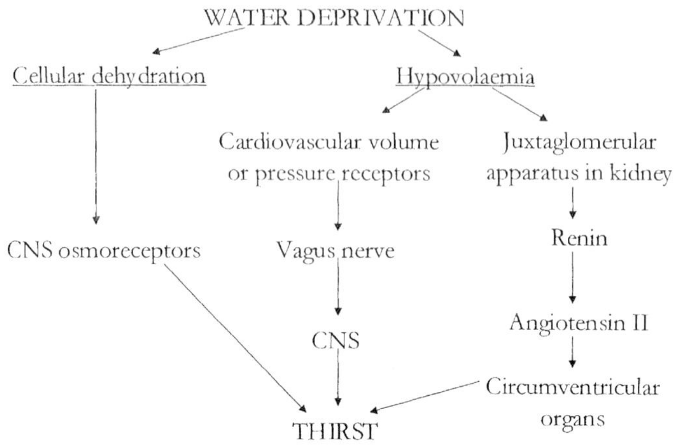

Figure 3. *A summary of pathways contributing to initiation of drinking following water deprivation.*

pre-absorptive factors implies some kind of metering of ingested water, which enables intake to closely match deficit. The rapid inhibitory effect of fluid ingestion extends beyond thirst to AVP secretion, as indicated by studies using water-deprived dogs (Thrasher *et al.*, 1981), which implied a key role here for the swallowing component of fluid consumption. In rats the oropharyngeal phase appears less important and the rapid inhibition of AVP secretion on drinking is apparently mediated by a visceral signal of osmotic dilution. The osmoreceptors or sodium receptors responsible are apparently in the intestines and/or hepatic-portal vein, which is an ideal location to provide 'anticipatory' information before water or osmolytes have entered the general circulation (Huang *et al.*, 2000). These peripheral receptors accordingly provide modulatory input, complementing the well-established negative feedback system provided by the osmoreceptors of the basal forebrain.

4. Integration of regulatory mechanisms

4.1 *Cellular dehydration, volaemic stimuli and regulatory hormones*

As described above cellular dehydration, as detected by osmoreceptors in the hypothalamic region, is the major stimulus regulating ADH secretion in mammals. Increased plasma and CSF osmolality leads to proportionate rise in firing rate of the magnocellular neurons of the supraoptic and paraventricular hypothalamic nuclei and increased neurohypophysial secretion of vasopressin (Robertson *et al.*, 1976; Wells and Balment, 1992). Volaemic stimuli from cardiovascular pressure and stretch receptors provide further modulatory input (Dunn *et al.*, 1973). By contrast, the control of renin secretion and thus ANG II formation, and indirectly aldosterone, is dominated by volaemic stimuli detected by cardiovascular and renal pressure and stretch receptors. Renin secretion from the afferent glomerular arterioles acts upon the blood-borne substrate angiotensinogen to generate the inactive decapeptide angiotensin I. Converting enzyme within the kidney and vasculature (particularly lungs) enables rapid cleavage of angiotensin I to the active octapeptide form angiotensin II (ANG II). A fall in blood pressure and/or vascular volume rapidly leads to a rise in circulating ANG II. Here further modulation of RAS activity relates to the sodium status of the animal, reflecting the monitoring role of the macula densa segment of the renal tubule, another component of the mammalian JG apparatus, which is sensitive to changes in tubular fluid Na. Regulation of the other major factor affecting fluid balance, ANP, is again dominated by volaemic stimuli, with cardiac secretion particularly sensitive to atrial stretch. Increased atrial stretch, indicative of elevated vascular volume and venous return leads to raised ANP secretion and resultant renal diuresis and natriuresis. Clearly during dehydration this pathway is greatly inhibited to support renal conservation of water.

The interaction and integration of cellular dehydration and hypo-volaemic stimuli regulatory inputs to renal management of water excretion are summarized in *Figure 4*.

4.2 *Interaction between central and peripheral regulatory systems*

As indicated above, in recent years it has become evident that the major regulatory peptide hormone systems (AVP, ANP, RAS), controlling renal water excretion, are also present within the CNS. Receptors for these peptides are found in the circumventricular organ region of the brain, which potentially can receive ligand input from both vascular and CSF/neuronal synapse sources (Reid, 1984). Actions of these peptides centrally can influence thirst and drinking behaviour (ANP profoundly

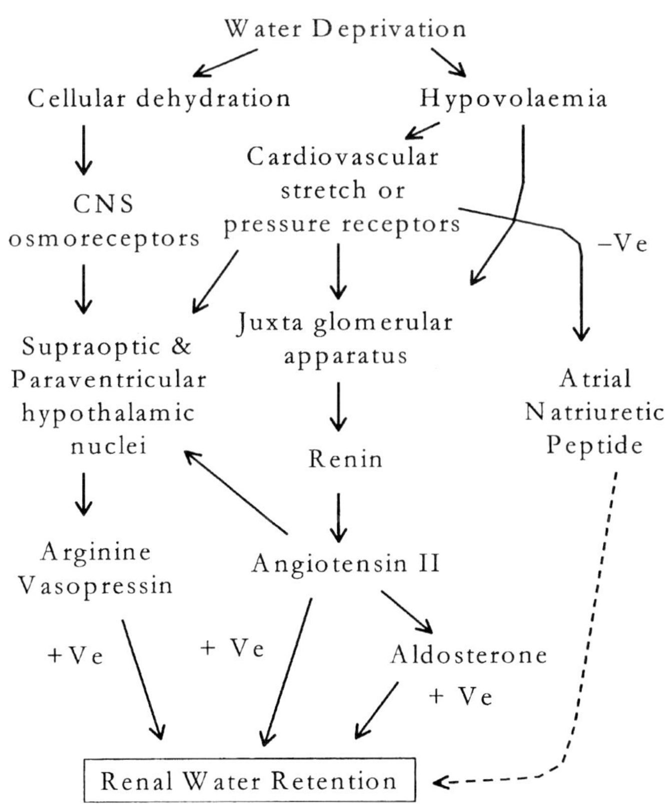

Figure 4. *Factors contributing to renal water conservation during water deprivation*

inhibits ANG II stimulated drinking (Thornton and Baldwin, 1988)), and neurohypophysial secretion of AVP (ANG II stimulates while ANP inhibits AVP secretion). Additionally these peptides can modulate regulation of systemic blood pressure (Abu-Amarah and Balment, 1999) and through altered renal nerve activity influence renal excretion of salt and water. This type of central action for ANG II is illustrated by the data shown in *Table 2*, taken from Wells and Balment (1990a).

Table 2 presents measures of urine flow and renal sodium excretion in an anaesthetized rat preparation during 1 hour of intracerebroventricular (ICV) infusion of either isotonic saline (1 μl min^{-1} vehicle) or ANG II (0.1 ng min^{-1}) with or without accompanying ANG II receptor blocker Saralasin (10 ng min^{-1}). By comparison with saline control an ANG II administration produces a significant reduction in urine flow and increase in sodium excretion, effects which are blocked by coincident infusion of Saralasin.

There is the potential for a complex interplay between these peripherally and centrally generated regulatory peptide systems. Currently, however, there is little evidence for direct cross-talk between them. This is illustrated by the data from Wells and Balment (1990b) in *Table 3*. By comparison with normal Long Evans rats, CSF ANG II levels are lower in the vasopressin-deficient Brattleboro rat, in association with their elevated peripheral plasma ANG II (165.4 ± 35 vs 70.3 ± 10.2 pg ml^{-1}, $n = 8$ for both). Also notable in *Table 3* is the raised plasma osmolality in Brattleboro relative to Long Evans rats, which is reflected in the higher CSF osmolality for Brattleboro rats. Interestingly, the depressed CSF ANG II level in Brattleboro rats persists even after 40 min intravenous converting enzyme inhibitor (70 μg min^{-1} Captopril) infusion to lower plasma ANG II levels. Similarly, acute elevation of plasma ANG II levels in Long Evans rats, by intravenous (0.5 μg min^{-1}) ANG II infusion, had no effect on CSF ANG II concentration in these animals. In a reciprocal study of normal rats in which central ANG II levels were raised by ICV 0.1 ng min^{-1} ANG II infusion for 1 hour, there was again no change in plasma ANG II concentration by comparison with vehicle infused controls (57.2 ± 8.3 vs 52.2 ± 4.7 pmol l^{-1}, $n = 7$). Accordingly, we have little support for direct cross-talk between peripheral and central RASs, at least in the short time-frame of these studies.

Nonetheless, it is apparent that some disorders of body fluid homeostasis are perhaps associated with disturbance of the normal balance between peripheral and central environments. A standard way in which cellular dehydration can be achieved without perturbing vascular volume is the intra-peritoneal (i.p.) injection of hypertonic saline. Data from such an experiment in anaesthetized rats are shown in *Figure*

Table 2. *Renal excretion during 1 hr ICV infusion of saline or angiotensin II in anaesthetized rat*

	Saline	Angiotensin II	Angiotensin II + Saralasin	Saralasin
Urine flow (μl min^{-1})	171 ± 10	121 ± 10	168 ± 12	187 ± 7
Na+ output (fmol min^{-1})	14.1 ± 1.0	21.8 ± 1.8	17.2 ± 1.3	15.0 ± 0.9

The values are mean ± SE for $n = 7$.

Table 3. *CSF and plasma composition following I.V. infusion of isotonic saline or angiotensin II in Long Evans rats, or saline or Captopril (converting enzyme inhibitor) in Brattleboro rats*

	Long Evans		Brattleboro	
	Saline	Angiotensin II	Saline	Captopril
CSF osmolality (mOsm kg^{-1})	298.4 ± 0.6	298.6 ± 0.9	304.7 ± 1.1	306.0 ± 1.7
CSF Ang II (pmol l^{-1})	49.1 ± 1.3	48.4 ±1.9	44.3 ± 1.7	41.3 ± 0.6
Plasma osmolality (mOsm kg^{-1})	292.0 ± 0.9	293.7 ± 0.8	298.4 ± 1.6	300.6 ± 1.4

Values are mean ± SE for n = 7.

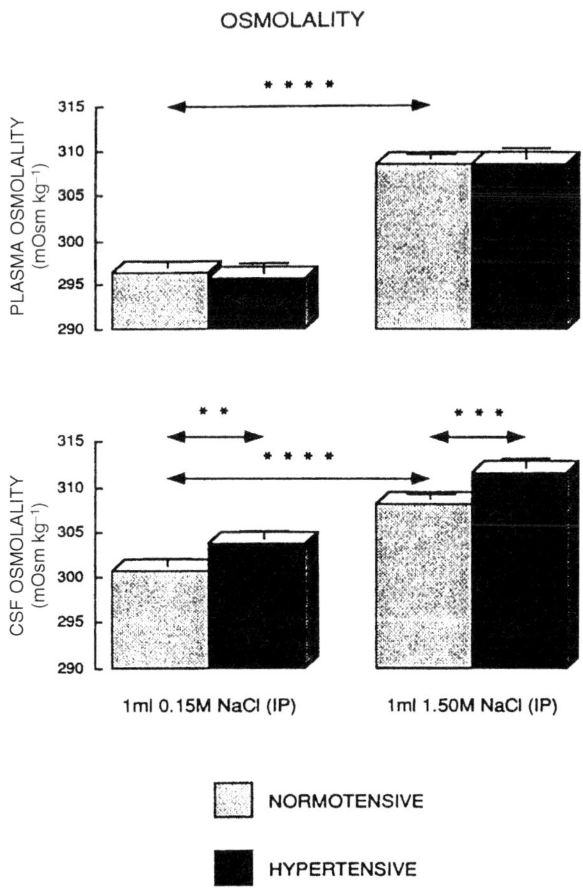

Figure 5. *Changes in CSF and plasma osmolality in normotensive and hypertensive rats (New Zealand strain) 20 min following intraperitoneal (I.P.) injection of isotonic (0.15 NaCl) or hypertonic (1.5 NaCl) saline. Statistical comparisons and levels of significance are indicated* **P < *0.01*,***P<0.005, ****P < *0.001*.

5, which presents measurements from normotensive and spontaneously hypertensive rats of the New Zealand strain (Wells and Balment, 1991).

Notable is the comparable plasma osmolalities for the two strains, 20 min after i.p. isotonic (0.15 M NaCl) saline injection. Despite this however the hypertensive rats exhibited elevated CSF osmolality relative to their normotensive controls, which was associated with elevated plasma AVP concentration (14.0 ± 3.2 vs 4.9 ± 0.9 pg ml^{-1}, $n = 6$ and 7, respectively). Following hypertonic saline (1.5 M NaCl) injection both groups of rats showed an expected and similar rise in plasma osmolality. This was reflected in a rise in CSF osmolality but again the hypertensive rats sustained higher CSF osmolality relative to the normotensive controls. The rise in CSF osmolality was accompanied by a rise in plasma AVP concentration in normotensive rats to 17.9 ± 2.2 pg ml^{-1} ($n = 6$). By contrast, this further rise in CSF osmolality in hypertensive animals was associated with a modest, non-significant rise in plasma AVP to 19.4 ± 1.6 pg ml^{-1} ($n = 7$). It appears that the normal relationship between plasma and CSF osmolality and its influence on AVP secretion is disturbed in this genetic model of hypertension. The development of hypertension, which reflects altered vascular fluid compartment regulation would, therefore, appear to potentially involve an abnormal AVP axis in this strain. Indeed, introduction of the AVP-deficient gene into this strain, through crossing with vasopressin-deficient Brattleboro rats, did lead to a significant reduction in the arterial blood pressure achieved in the adult hypertensive animals (Ashton and Balment, 1989).

5. Integration of central and peripheral mechanisms to maintain water balance

In dehydration cerebral osmoreceptors and sodium sensors, together with extracellular fluid volume detectors and pressure sensors, are involved in regulation of vasopressin secretion, thirst, and renal excretion of sodium and water (McKinley *et al.*, 1985). Hypo-volaemia is perhaps a more immediate threat to life in mammals than is cellular dehydration and mechanisms for the preservation and restoration of circulatory volume are critical for survival. The temporary amelioration of blood loss afforded by the immediate reflex responses of the heart and blood vessels is rapidly consolidated by redistribution of interstitial fluid and by increased water drinking, followed by increased sodium intake. The threshold for eliciting hypo-volaemic thirst is 8–10% blood loss compared with a threshold of only 1–2% loss of cell water for osmotic thirst. The higher threshold for hypo-volaemic thirst largely reflects the great range in blood flows that essential organs can accommodate and the capacity for substantive redistribution of cardiac output. Further, the large volume of interstitial fluid can act as a buffer to be mobilized to maintain circulation as the need arises. When hypo-volaemia persists as a threat to survival, the drinking evoked is as profound as that seen with equivalent cellular dehydration (Fitzsimons, 1998).

Activation of the RAS in hypo-volaemia exemplifies the ability to co-ordinate a number of responses (thirst, sodium appetite, haemodynamic, renal, vasopressin and aldosterone secretion) to repair loss of circulating volume. As discussed above the responses to cellular dehydration via peripheral and central osmoreceptors/sodium receptors also permit co-ordinated activation of thirst mechanisms and, via ADH secretion, renal conservation of water. During water deprivation, as cellular dehydration and hypo-volaemia both develop these separate complex pathways

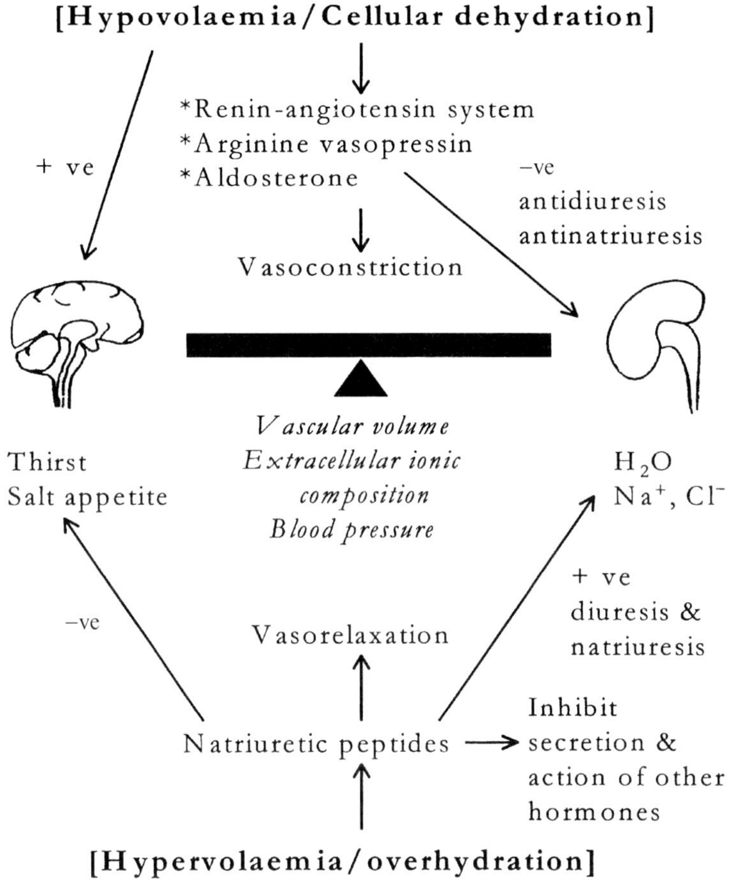

Figure 6. *Diagrammatic summary of the two major balancing regulatory pathways securing maintenance of water and ion homeostasis, and blood pressure.*

clearly interact additively if not synergistically, at the levels of thirst, AVP secretion, cardiovascular responses and renal water and salt conservation, to secure effective rehydration and maintenance of body fluid composition and blood pressure.

The contrasting situation of overhydration and volume expansion is associated with activation of the cardiac natriuretic peptide system. In contrast to fish, where raised plasma osmolality is more important (Takei, 2000), for mammals an increase in vascular volume and atrial stretch is the major stimulus. Atrial natriuretic peptides acting centrally inhibit drinking behaviour and sodium appetite (Antunes-Rodrigues *et al.*, 1985; Weisinger *et al.*, 1992). These central activities are complemented by the effects of circulating natriuretic peptides to inhibit sympathetic nervous activity, vasopressin secretion, renal renin release and aldosterone secretion, leading to a fall in blood pressure and increased urinary loss of sodium and water.

Accordingly, in mammals we have two major balancing pathways to support the maintenance of body fluids; each involves a combination of central and peripheral regulatory factors and effectors and each is dependent upon co-ordinating central and peripheral (cardiovascular, renal) mechanisms. The two pathways separately support either water conservation and repletion in dehydration, or excess water elimination and protection against overhydration. These opposing pathways are diagrammatically summarized in *Figure 6*. The relative activity of the two pathways is in turn dependent upon the extent of any deviation from the optimum set point for each of the principle factors (vascular volume, extra-cellular fluid, ionic composition, blood pressure) being maintained. There is also clear cross-talk between the two pathways. This is apparently largely based upon the inhibitory action of natriuretic peptide at the levels of thirst, AVP secretion, aldosterone secretion, vascular tone, renal conservation of sodium and water. The relative importance of these two systems and the balance of their interactions varies between species perhaps in relation to the water availability and threat of dehydration in their dominant environment.

References

Abu-Amarah, I. and Balment, R.J. (1999) Vascular, renal and endocrine responses to low dose ANP in the fluid balanced New Zealand genetically hypertensive rat with and without endogenous arginine vasopressin. *Can. J. Physiol. Pharmacol.* **77**: 102–110.

Acher, R. (1996) Molecular evolution of fish neurohypophysial hormones, neutral and selective evolutionary mechanisms. *Gen. Comp. Endocrinol.* **102**: 157–172.

Adolf, E.F. (1950) Thirst and its inhibitions in the stomach. *Am. J. Physiol.* **161**: 374–386.

Andersson, B. (1978) Regulation of water intake. *Physiol. Ren.* **58**: 582–603.

Andersson, B., Eriksson, L., Fernandez, O., Kolmodin, C.A. and Oltner, R. (1972) Centrally mediated effects of sodium and angiotensin II on arterial blood pressure and fluid balance. *Acta. Physiol. Scand.* **85**: 398–407.

Antunes-Rodrigues, J., McCann, S.M., Rogers, L.C. and Samson, W.K. (1985) Atrial natriuretic factor inhibits dehydration – and angiotensin II-induced water intake in the conscious, unrestrained rat. *Proc. Natl Acad. Sci. USA.* **82**: 8720–8723.

Ashton, N. and Balment, R.J. (1989) Blood pressure and renal function in a novel vasopressin-deficient, genetically hypertensive rat strain. *J. Physiol.* **410**: 21–34.

Balment, R.J., Henderson, I.W., Chester-Jones, I. and Mosley, W. (1977) Water and electrolyte balance in adrenalectomized rats with diabetes insipidus (Brattleboro strain) given antidiuretic hormone. *Gen. Comp. Endocrinol.* **33**: 428–433.

Bannermann, B., Faxe, K. and Ganten, D. (1993) The renin–angiotensin system in the brain: an update. *Regul. Pept.* **46**: 487–509.

Burgess, W.J. and Balment, R.J. (1993) Effect of atrial natriuretic peptide (ANP) on vasopressin-stimulant cAMP accummulation in microdissected rat nephron segments. *J. Endocrinol.* **137**: Suppl. 155.

de Rouffignac, C., di Stefano, A., Wittner, M., Roiscel, N. and Elalouf, J.M. (1991) Consequences of differential effects of ADH and other peptide hormones on thick ascending limb of mammalian kidney. *Am. J. Physiol.* **260**: R1023–R1035.

Dunn, F.L., Brennan, T.J., Nelson, A.F. and Robertson, G.L. (1973) The role of blood osmolality and volume in regulating vasopressin secretion in the rat. *J. Clin. Invest.* **52**: 3212–3219.

Emmeluth, C., Drummer, C., Gerzer, R. and Bie, P. (1992). Roles of cephalic Na^+ concentration and urodilatin in control of renal Na^+ excretion. *Am. J. Physiol.* **262 (3Pt 2)**: F513–F516.

Fitzsimons, J.T. (1969) The role of renal thirst factor in drinking induced by extracellular stimuli. *J. Physiol.* **201**: 349–368.

Fitzsimons, J.T. (1980) Peptides, thirst and sodium appetite. *Proc. R. Soc. Lond. B. Biol Sci.* **210**: 165–182.

Fitzsimons, J.T. (1998) Angiotensin, thirst and sodium appetite. *Physiol Rev.* **78**: 583–686.

Fitzsimons, J.T. and Simons, B.J. (1969). The effects on drinking in the rat of intravenous infusion of angiotensin, given alone or in combination with other stimuli of thirst. *J. Physiol.* **203:** 45–57.

Forssmann, W-G. (1995) Urodilation (Ulartide, INN): A renal natriuretic peptide; Molecular Biology and clinical importance. *Nephron.* **69**: 211–222.

Gilman, A. (1937) The relations between blood osmotic pressure, fluid distribution and voluntary water intake. *Am. J. Physiol.* **120**: 323–328.

Gutkowska, J., Antunes-Rodrigues, J. and McCann, S.M. (1997) Atrial natriuretic peptide in brain and pituitary gland. *Physiol Rev.* **77**: 465–515.

Hall, J.E., Guyton, A.C. and Mizelle, H.L. (1990) Role of the renin–angiotensin system in control of sodium excretion and arterial pressure. *Acta. Physiol. Scand.* **139**: Supplementum 591, 48–62.

Hillyard, S.D., Hoff, KvS. and Propper C. (1998) The water absorption response: a behavioural assay for physiological processes in terrestrial amphibians. *Physiol. Zool.* **71**: 127–138.

Huang, W., Sved, A.F. and Stricker, E.M. (2000) Water ingestion provides an early signal inhibiting osmotically stimulated vasopressin secretion in rats. *Am. J. Physiol.* **279**: R756–R760.

Imura, H., Nakao, K. and Itoh, H. (1992) The natriuretic peptide system in the brain: implications in the central control of cardiovascular and neuroendocrine functions. *Front Neuroendocrinol.* **13**: 217–249.

McKinley, M.J. (1991) Osmoreceptors for thirst. In: *Thirst: Physiological and Psychological Aspects* (eds D.J. Ramsay and D.A. Booth). Springer, London, pp. 77–90.

McKinley, M.J., Congiu, M., Denton, D.A., Lichardus, B., Park, R.G., Tarjan, E. and Weisinger, R.S. (1985) In: *Vasopressin* (ed. R.W. Schrier). Raven Press, New York, pp. 299–309.

Nielson, S. and Agre, P. (1995) The aquaporin family of water channels in kidney. *Kidney Int.* **48**: 1057–1068.

Oliver, J.A., Henderson, I.W., McKeever, A. and Balment, R.J. (1982) The renin–angiotensin system in the absence of antidiuretic hormone. *Annals New York Acad. Sci.* **394**: 275–277.

Pang, P.K.T., Farspan, P.B. and Sawyer, W.H. (1983) Evolution of neurohypophysial hormone action in vertebrates. *Am. Zool.* **23**: 655–662.

Phillips, M.I, Heininger, F. and Tofbolo, S. (1996) The role of brain angiotensin in thirst and AVP release induced by haemorrhage. *Regul. Pept.* **66**: 3–11.

Reid, I.A. (1984) Actions of angiotensin II on the brain: mechanisms and physiologic role. *Am. J. Physiol.* **246**: F533–F543.

Robertson, G.L., Shelton, R.L. and Athar, S. (1976) The osmoregulation of vasopressin. *Kidney Int.* **10**: 25–37.

Rolls, B.J. and McFarland, D.J. (1973) Hydration releases inhibition of feeding produced by intracranial angiotensin. *Physiol. Behav.* **11**: 881–884.

Rolls, B.J. and Rolls, E.T. (1982) *Thirst – (Problems in the behavioural sciences).* Cambridge University Press, Cambridge.

Ruskoaho, H. (1992) Atrial natriuretic peptide: synthesis, release and metabolism. *Pharmac. Rev.* **44**: 479–602.

Saaredra, J.M. (1992) Brain and pituitary angiotensin. *Endocrinol. Rev.* **13**: 329–380.

Stricker, E.M. (1981) Thirst and sodium appetite after colloid treatment in rats. *J. Comp. Physiol.* **95**: 1–25.

Takei, Y. (2000) Comparative physiology of body fluid regulation in vertebrates with special reference to thirst regulation. *Japanese J. Physiol.* **50**: 171–186.

Thornton, S.N. and Baldwin, B.A. (1988) Centrally injected atrial natriuretic peptide inhibits angiotensin – and osmotically induced drinking in pigs. *Q.J. Exp. Physiol.* **73**: 1009–1012.

Thrasher, T.N., Nistal-Herrera, J.F., Keil, L.C. and Ramsay, D.J. (1981) Satiety and inhibition of vasopressin secretion after drinking in dehydrated dogs. *Am. J. Physiol.* **240**: E394–E401.

Wargent, E.T. and Balment, R.J. (1997) AVP–steroid interactions in the rat inner medullary collecting duct. *J. Endocrinol.* **155**: Suppl. OC32.

Weisinger, R.S., Blair-West, J.R., Denton, D.A. and Tarjan, F. (1992) Central administration of atrial natriuretic peptide suppresses sodium and water intake in sheep. *Brain Res.* **579**: 113–118.

Wells, T. and Balment, R.J. (1990a) The renal effects of central angiotensin II (Ang II) administration in the anaesthetised rat are inhibited by saralasin. *J. Physiol.* **430**: 148P.

Wells, T. and Balment, R.J. (1990b) Interactions of central and peripheral renin–angiotensin systems (RAS) in the anaesthetised rat. *J. Endocr.* **127**: Suppl. 107.

Wells, T. and Balment, R.J. (1991) Vasopressin secretion, plasma and cerebrospinal fluid (CSF) composition following peripheral osmotic stimulation in anaesthetised normotensive and hypertensive rats. *J. Physiol.* **434**: 79P.

Wells, T. and Balment, R.J. (1992) The effect of chronic and acute changes in plasma composition on vasopressin secretion and cerebrospinal fluid in the rat. *Acta Endocrinol.* **127**: 174–178.

11

Organismal handling of water balance in vertebrates

R. Acher and J. Chauvet

1. Introduction

Water, the 'ash of hydrogen', is the main component of living beings representing about 70% of the body weight in vertebrates and reaching 90–95% in invertebrates and plants. If the proportions are evaluated on a molar rather than a weight basis, roughly 99.9% of the molecules making up the body are water molecules. Although water as ice crystals antedates the accretion of earth, the liquid form, requiring appropriate conditions of temperature and pressure, may have been associated with the appearance of life, some 3.6–3.8 billion (10^9) years ago (Mojzsis *et al.*, 1996; Rosing, 1999), represented in all likelihood by cyanobacterium-like microorganisms (Schopf, 1993). Life was exclusively prokaryotic for about 2 billion years (Doolittle *et al.*, 1996) and prokaryotic cells had to set up primordial osmoregulatory mechanisms to face medium changes.

With the rise of unicellular eukaryotes and formation of organelles such as endoplasmic reticulum, Golgi apparatus, mitochondria and chloroplasts, water and osmolyte homeostases had to be specifically secured in each cell compartment. Finally, in multicellular eukaryotes with extra-cellular fluids, a particular osmoregulation of the circulating milieu was established under hormonal control. In vertebrates, therefore, water and osmolyte handling concerns three levels: that of the body fluids (blood, volaemia and osmolality), of the cytoplasmic compartment and of the intra-cellular organelles. Our knowledge of the mechanisms involved in regulation of intra-cellular compartments is still very fragmentary; this study will mainly concern the regulation of water and salt balances of the extra-cellular fluids in vertebrates, with an emphasis on neurohypophysial control (Acher, 1995). However, some general information on osmoregulation in prokaryotes is required to understand the molecular responses of animal cells to osmotic stress.

Osmoregulation and Drinking in Vertebrates, edited by N. Hazon & G. Flik.
© 2002 BIOS Scientific Publishers Ltd, Oxford

2. The primordial osmoregulation in prokaryotic and eukaryotic cells

Cells are enclosed by a membrane that is readily permeated by water but does not allow free diffusion of most solutes. Changes in external osmolality result, therefore, in osmotic stress because of unequal rates of movement of water and solutes across the membrane. Cells respond to osmotic stress by regulatory compensation of changes in cell volume, water content and intra-cellular solute concentrations. They adjust the levels of compatible osmolytes, i.e. small organic solutes such as glycine betaine or *myo*-inositol, that are generally accumulated by transporters or enzymes, many of which are transcriptionally regulated (Kültz and Csonka, 1999).

Of fundamental importance is the homeostasis of intra-cellular inorganic ion concentrations. Most cells maintain lower intra-cellular Na^+ and higher K^+ concentrations than those present in the external milieu, and the resulting gradients create the driving-force for energy-requiring processes. Osmotic stress disturbs the intra-cellular ionic milieu and interferes with cell function. Restoration to normal implies osmoregulatory responses including the accumulation of compatible osmolytes. In cells in which a cytoplasmic membrane is surrounded by a cell wall as in bacteria and plants, the osmolality generated by compatible osmolytes can exceed the external osmolality and thus promotes influx of water and generates an outwardly directed pressure supported by the cell wall; this so-called *turgor* could be an osmoregulatory signal (Kültz and Csonka, 1999). In *Lactococcus lactis,* an osmoregulated ABC transporter (OpuA) has been identified that responds to water stress. This glycine betaine transport system consists of an ATP binding and hydrolysing subunit (OpuAA) and a protein (OpuABC) that contains both the translocator and the substrate-binding domain. The components of OpuA have been purified and incorporated into liposomes with an ATP-generating system in the vesicle lumen. A trans-membrane osmotic gradient of both ionic and non-ionic compounds osmotically activates OpuA in this proteoliposomal system; hypo-osmotic medium inhibited the basic activity of the system. OpuA can act both as osmosensor and osmoregulator (Van der Heide and Poolman, 2000).

In animal cells—that do not possess cell walls and cannot maintain an osmotic pressure difference between internal and external milieu—the phenomenon of accumulation of organic compounds in response to a hyper-osmotic stress to compensate for down-regulated intra-cellular Na^+ and K^+ levels has been preserved. Cells survive in a hypertonic environment by increasing the transcription of genes whose products catalyse cellular accumulation of compatible osmolytes. The mammalian kidney medulla is normally hypertonic because of the urine concentration mechanism (Kültz and Csonka, 1999). Cellular accumulation of compatible osmolytes in the renal medulla is catalysed by a sodium/*myo*-inositol-cotransporter (SMIT), a sodium/chloride/betaine-cotransporter and aldose reductase (key in synthesis of sorbitol). Specific inhibition of SMIT *in vivo* results in necrotic injury of the renal medulla and subsequent renal failure (Kültz and Csonka, 1999). Tonicity-responsive enhancers (TonE) play a key role in hypertonicity-induced transcriptional stimulation of SMIT, of the sodium/chloride/betaine-cotransporter and aldose reductase. The cDNA cloning of human TonE binding protein (TonEBP), a transcription factor that stimulates transcription through its binding to TonE sequence via a Rel-like DNA binding domain, has been carried out (Miyakawa *et al.*, 1999). Western blot and immunohistochemical analyses of cells cultured in hypertonic medium reveal that hypertonicity elicits slow activation of TonEBP, which is the result of an increase in the TonEBP amount and translocation to the nucleus (Miyakawa *et al.*, 1999). TonEBP is responsible for transcriptional activation of the osmoprotective genes encoding the glycine betain transporter, BGT1, and the *myo*-inositol transporter, SMIT, during hyper-osmotic stress in mammalian cells. Interestingly, this protein bears no similarity to transcription factors that regulate osmoprotective genes in prokaryotes, plant cells or yeasts.

3. The osmotic balance in vertebrates

Water homeostasis, i.e. the maintenance of a stable water volume, depends on the regulated balance between intake and excretion of water and osmolytes such as salts or urea. In vertebrates, water homeostasis is controlled by at least three hormonal systems: the hypothalamo-neurohypophysial system and the renin–angiotensin–aldosterone system, both acting for the recovery of water and sodium from urine, and the atrial natriuretic factor system operating the opposite way. We will address here only the first-mentioned regulatory system; the other system is referred to in other chapters in this volume.

The neuro-endocrine reflex involved in the hypothalamo-neurohypophysial regulation implies an afferent neural branch from sensors, such as osmoreceptors for osmolality (Bourque *et al.*, 1994) and baroreceptors for volaemia (Thrasher, 1994), to the hypothalamus and an efferent neuro-endocrine branch from hormone secreting neurons to the target cells with their receptors and effector mechanisms. The latter branch comprises a cascade of events that can now be described in molecular terms as a succession of protein–protein specific interactions *(Figure 1)*. From precursor synthesis and processing into the mature mediator(s) to receptor recognition and signal transduction to effectors, the shift in the conformation or three-dimensional structure of proteins (conformation itself being largely determined by the spontaneous folding of the polypeptide chain), appears to be the crucial mechanism to ensure the precise physiological response (Acher and Chauvet, 1995).

Water must be distributed over a number of body compartments. In man about 55% of total body water is intra-cellular and 45% extra-cellular, the water in the blood plasma represents 7.5% (Quan and Cogan, 1993). The hormonal control of water permeability of the plasma membrane and different intra-cellular membrane compartments in differentiated cells remains largely to be investigated.

One of the major problems encountered by vertebrates during the colonization of the earth was the maintenance of volume and composition of the body fluids in a more or less stable state. All vertebrates, with the exception of hagfish (Cyclostomata) that are iso-osmotic to the sea (see Rankin, this volume), keep the osmotic pressure of the blood at an essentially fixed value, regardless of the medium in which they live. Bony vertebrates in fresh water (FW), marine environments or on land have a blood osmolality in the range of 250–350 mOsm kg^{-1} H_2O using mainly salts as osmolyte. In contrast, cartilaginous fish adjust their blood osmolality slightly above that of the medium to around 1050 mOsm kg^{-1} for marine species, using the uncharged urea as osmolyte (Bentley, 1972). The latter ability was preserved in the *Crossopterygii* (lungfish, coelacanth), the bony fish that are at the basis of the evolutionary transition to the terrestrial tetrapods. The same mechanism is also found in

The osmoregulatory neuroendocrine reflex

I/	**Afferent neural branch**	
	1. Osmoreceptors:	Osmolality
	2. Baroreceptors:	Volaemia
II/	**Efferent endocrine branch**	
	1. *Precursors:*	Hypothalamic neurons
	Biosynthesis	
	Processing	
	2. *Mediators:*	Hypothalamic neurons
	Storage	
	Secretion	
	3. *Receptors:*	Basal membrane
	Biosynthesis	
	Translocation	
	4. *Effectors:*	Apical membrane
	Water channel	Aquaporin 2
	Urea transporter	UT1
	Sodium channel	EnaC
	Potassium channel	ROM K1
	Chloride channel	CFTR

Figure 1. *The cascade of molecular events underlying the neuroendocrine reflex controlling hydro-osmotic homeostasis.*

some amphibia that live in arid countries and face dehydration (Katz, 1990). Conservation of urea as osmolyte implies that an animal must have conserved the hepatic enzymes for urea production from arginine as well as the renal urea reabsorption mechanism with its active transporters.

Among terrestrial vertebrates, only mammals and birds secrete a urine that is hypertonic to the blood plasma and so use much less water for renal solute excretion than amphibians or reptiles. They can do so because of the presence of the Henle's loop in the nephron. However, neurohypophysial hormone-regulated antidiuresis, by decreased glomerular filtration rate or tubular reabsorption, is found in all tetrapods and must be considered an early adaptation to terrestrial life. However, morphological, physiological and endocrinological modifications were necessary to allow the organisms to adapt to the three main types of environments on earth: FW, sea and air.

4. The cascade of protein–protein interactions underlying the neurohypophysial control of osmoregulation

Any neuro-endocrine regulatory process can be defined as a cascade of protein–protein interactions. In the neuro-secretory cell we find preprohormone protein gene expression, multi-step precursor processing, vesicular transport of the neurohormone and targeted exocytosis. In the target cell, the cascade includes specific mediator–receptor recognition, signal transduction, cytosolic enzyme activation, effector mobilization and for instance aquaporin containing vesicle transport to the apical membrane (see 4.3). In the case of the mammalian neurohypophysial hormone vasopressin that controls water reabsorption from the inner medullary collecting duct through epithelial principal cells, the cascade of interactions has now been clearly described, virtually all the protein components have been cloned and many have been characterized even to the level of their three-dimensional structure.

4.1 *Preprovasopressin: gene expression, precursor transport and processing*

The genes for preprovasopressin and preprooxytocin, the two mammalian neurohypophysial precursors, were identified in man (Sausville *et al.*, 1985), rat (Ivell and Richter, 1984), ox (Ruppert *et al.*, 1984) and mice (Hara *et al.*, 1990). They have the same organization encoding for a three- or a four-domain protein, namely an 18–20 amino acid residue signal peptide, the nonapeptide hormone, a 93–95 amino acid residue neurophysin and in the case of preprovasopressin a 39 amino acid residue copeptin. In non-mammalian vertebrates, preprovasotocin replaces preprovasopressin but the four-domain structure is preserved. The processing occurs during the transport of the precursor from its entrance into the rough endoplasmic reticulum (pH 7) to the cis-, median- and trans-compartments of the Golgi system and finally the secretory granules (pH 5.0) in which the hormone matures. A cascade of four enzymes, likely associated in a membrane-anchored machinery, is operating: 1) a calcium-dependent Lys-Arg endopeptidase belonging to the subtilisin-like family (Proconvertases PC1, PC2); 2) a zinc carboxypeptidase B-like exoprotease, carboxypeptidase E, activated by cobalt, that removes Lys-Arg residues; 3) a peptidyl-glycine monooxygenase that oxidizes the C-terminal glycine into hydroxyglycine; 4) a lyase that splits hydroxyglycine into an amide group for the penultimate glycine and glyoxylic acid (Acher and Chauvet, 1995). Mature, fully-processed carboxyamidated vasopressin displays the bioactive conformation recognized by the specific receptors. Within the storage granules the neurohormones proper, vasopressin-associated and oxytocin-associated neurophysins and copeptin as well as the enzymes involved in the processing are found. Targeting of granules to secretion sites of the end nerve membrane is likely directed through a synaptobrevin-syntaxin system as in synaptic vesicles; secretion occurs by exocytosis (Fernandez- Chacon and Südhof, 1999). Secretion is triggered by afferent signals coming from osmo- and baroreceptors.

4.2. *Vasopressin receptor signalling, GTPase (G) protein, adenylate cyclase and protein kinase A*

In its target cells, the hydro-osmotic principal cells of the inner medullary collecting duct, binding of vasopressin occurs to the heptatransmembrane vasopressin receptor V2. Binding of the hormone to the extra-cellular neurohormone-binding domain of the receptor triggers a conformational change in the cytoplasmic Gs-protein binding domain (likely the third cytoplasmic loop of the receptor). This heterotrimeric G-protein dissociates into an alpha-subunit and beta-gamma-subunits. The alpha-subunit in turn associates with adenylate cyclase and activates this latter through GTP hydrolysis. Adenylate cyclase forms cAMP from ATP. Cyclic AMP links the regulatory subunit of protein kinase A with release of the catalytic subunit. This enzyme phosphorylates the five effectors that are under vasopressin control, viz. the water channel aquaporin 2 (AQP2), the urea transporter UT1, the amiloride-sensitive sodium channel ENaC (Matalon, 1999), the inward-rectifying potassium channel ROM-K1 (Wang *et al.*, 1997), and the chloride channel CFTR.

4.3. *Cytosol–apical membrane shuttle vesicle*

Quiescent effectors are located in cytosol, likely in vesicles termed *aggrephores*. Indeed, in such vesicles the six-transmembrane protein AQP2, the protein kinase A (PKA), a specific A kinase anchoring protein, AKAP, that controls phosphorylation of AQP2, and likely a phosphatase 2B are found. The phosphorylation of AQP2 induces the functional conformation. Vesicles are targeted to the apical membrane through kinesin motor proteins along microtubules (Setou *et al.*, 2000); other membrane proteins such as the couple synaptobrevin (VAMP2)-syntaxin ensure the fusion with apical membrane required for exocytosis (Klussmann *et al.*, 2000). Fusion of aggrephores with the apical membrane increases the number of water channels and therefore the water permeability of this membrane compartment. Water from the lumen of the collecting duct enters the cells through aquaporin 2, diffuses through the cytosol and exits the cell through AQPs 3 and 4. From the interstitium, water molecules enter the capillaries through AQP1 and reach the general circulation *(Figure 2)*. Dephosphorylation of AQP2 by phosphatase 2B, perhaps involving a phosphatase anchoring protein, triggers endocytosis and the return of AQP2 to the cytosol. Thus is desensitization of the target cell to vasopressin realized.

5. The unique urea-based osmoregulation in elasmobranchs

The species of the different vertebrate classes have evolved specific sets of osmoregulatory organs to face medium changes and to maintain their hydro-osmotic homeostasis *(Table 1)*. Cartilaginous fish were the first to develop a particular strategy

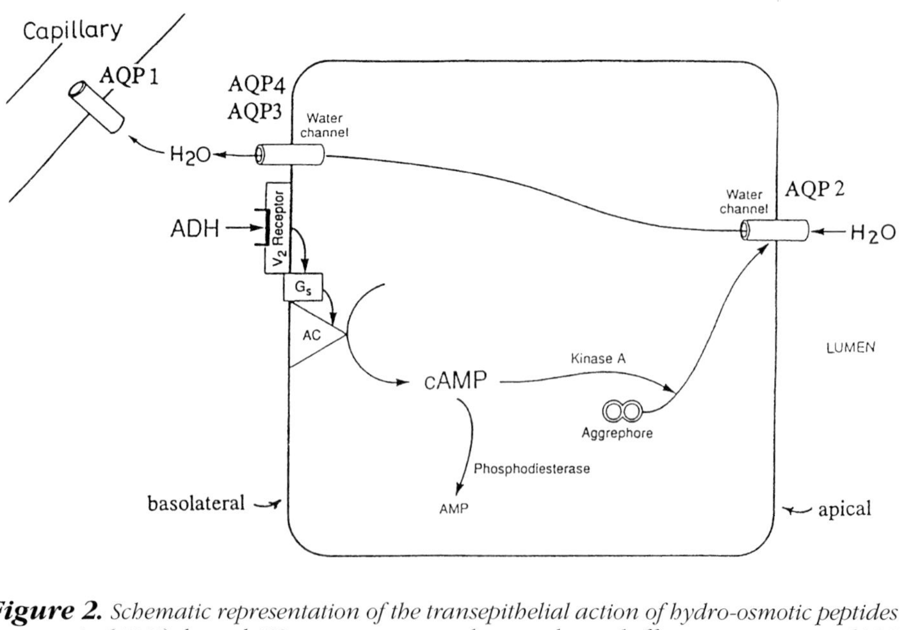

Figure 2. *Schematic representation of the transepithelial action of hydro-osmotic peptides vasopressin (ADH) through V2-type receptors and water channel effectors, aquaporins (AQP).*

Table 1. *The osmoregulatory organs specific for each class of vertebrates*

Cartilaginous fishes	Bony fishes	Amphibians	Reptiles	Birds	Mammals
Gut	Gut	Gut	Gut	Gut	Gut
Kidney	Kidney	Kidney	Kidney	Kidney	Kidney
Gills	Gills	Bladder	Nasal gland	Nasal gland	
Rectal gland		Skin			

in osmoregulation involving the organic osmolyte urea to increase the osmotic value of their blood and diminish the osmotic gradient between body and medium.

5.1 *The urea-based osmoregulatory system*

Whereas the main osmolyte in the blood plasma of bony vertebrates is NaCl, determining an osmolality in the range of 250–400 mOsm kg^{-1}, in marine cartilaginous fish NaCl contributes only 50% of the total osmolality; the remainder is contributed by urea (35–40%) and trimethylamine oxide (TMAO, 5–20%; Perlman and Goldstein, 1988; Shuttleworth, 1988). The total plasma osmotic concentration in marine chondrichtyans is usually in the range 1000–1100 mOsm l^{-1}, but always some 2–10% higher than that of the surrounding seawater.

The osmoregulatory system in Chondrichthyes involves four main osmoregulatory organs: gut, kidney, gills and the rectal gland. Water, salt and nutrient uptake occurs across the intestinal epithelium; water, salt and the nitrogen metabolites urea and TMAO are filtered into the nephron, where the filtrate is subsequently processed. In cartilaginous fish, the nephron is very complex and consists of 16 morphologically distinct segments (Lacy and Reale, 1995). About 75–85% of the filtered fluid is reabsorbed in the nephron and, together with solute reabsorption, this renders urine hypo-osmotic to plasma. On the other hand, 90–95% of the filtered urea and TMAO are reabsorbed to serve as important body fluid osmolytes (Perlman and Goldstein, 1988). In contrast to most teleosts in which ammonia, the major nitrogen end product, is excreted essentially through the gills, in elasmobranchs only 25% or less of the gill excretion of nitrogen is in the form of ammonia (Perlman and Goldstein, 1988).

In mammals, vasopressin has been shown to stimulate renal tubule water and urea reabsorption by increasing the number of water channels and urea transporters (Shayakul *et al.*, 1996) in the renal collecting duct cells (Shuttleworth, 1988). It is plausible that vasotocin, the counterpart of vasopressin in non-mammalian vertebrates, plays a similar role in elasmobranchs in which urea reabsorption is crucial for osmoregulation. The possible role of the specific neurohypophysial hormones found in cartilaginous fish has recently been investigated (Acher *et al.*, 1999).

5.2 *Neurohypophysial hormones of cartilaginous fish*

The class Chondrichthyes comprises the subclasses Holocephali (ratfish) and Elasmobranchii (rays, skates and sharks). The oldest geological record of Chondrichthyes dates back to the early Devonian some 450 million years ago. Chondrichthyes are

	1	2	3	4	5	6	7	8	9
Ancestral vasotocin:	**Cys**	**Tyr**	**Ile**	**Gln**	**Asn**	**Cys**	**Pro**	**Arg**	**Gly (NH_2)**
Vasopressin Lineage									
Vasotocin	–	–	–	–	–	–	–	–	–
Vasopressin	–	–	Phe	–	–	–	–	–	–
Lysipressin	–	–	Phe	–	–	–	–	Lys	–
Phenypressin	–	Phe	Phe	–	–	–	–	–	–
Oxytocin Lineage									
Isotocin	–	–	–	Ser	–	–	–	Ile	–
Mesotocin	–	–	–	–	–	–	–	Ile	–
Oxytocin	–	–	–	–	–	–	–	Leu	–
Glumitocin	–	–	–	Ser	–	–	–	Gln	–
Aspargtocin	–	–	–	Asn	–	–	–	Leu	–
Valitocin	–	–	–	–	–	–	–	Val	–
Asvatocin	–	–	–	Asn	–	–	–	Val	–
Phasvatocin	–	–	Phe	Asn	–	–	–	Val	–

Figure 3. *The two lineages of neurohypophysial hormones, the vasopressin-like and the oxytocin-like peptides, identified in vertebrates.*

assumed to have a marine origin and, with a few exceptions, all extant species live in a marine environment (Evans, 1993). In 1994, Nelson recognized about 846 species, including about 425 skates and rays, 350 sharks and 30 chimaeras (about 3.5% of all extant fish).

In the frame of our studies on fish neurohypophysial hormones (Acher, 1996; Acher *et al.*, 1997a), we have investigated the neurohypophysial hormones of species belonging to the two Chondrichthyan subclasses, viz. of four rays and of two dogfish (Elasmobranchii) and of the Pacific ratfish (Holocephali). As all non-mammalian vertebrates, cartilaginous fish possess vasotocin ([Ile^3]-vasopressin), a peptide preserved in the course of evolution from cyclostomes to birds, and replaced only in mammals by vasopressin. This remarkable evolutionary stability, during a period of roughly 450 million years, may find its origin in a common essential function. However, although the antidiuretic role of vasotocin is apparent in non-mammalian tetrapods, this is not the case in fish. When bony and cartilaginous fish are compared, a striking difference can be observed in the amount of vasotocin stored in the neurohypophysis. Vasotocin was initially believed to be absent in Chondrichthyes and the scarcity of this peptide in tissues of this group hampered purification. However, we have been able to identify vasotocin in all species we examined so far, using ion-exchange chromatography and specific pharmacological assays.

Thus all cartilaginous fish have vasotocin, albeit it stored in very small amounts in the neurohypophysis. The oxytocin-like hormones of Elasmobranchii differ from those characterized in bony vertebrates. Whereas the rays possess glumitocin ([Ser^4,Gln^8]-oxytocin), the sharks display both a duplication and diversification of this neurohormone. In *Squalus acanthias*, *aspargtocin* ([Asn^4]-oxytocin) and *valitocin* ([Val^8]-oxytocin) have been identified whereas in *Scyliorhinus caniculus*, *asvatocin* ([Asn^4,Val^8]-oxytocin) and *phasvatocin* ([Phe^3,Asn^4,Val^8]-oxytocin), have been found (Chauvet *et al.*, 1994). Finally and unexpectedly, oxytocin was characterized in the Pacific ratfish *(Figure 3)*.

Because of the large amounts of the oxytocin-like hormones in the pituitary gland of cartilaginous fish (compared to vasotocin), an important physiological function is likely and these peptides could possibly play a role in the reabsorption of urea. This hypothesis remains to be investigated.

6. Co-evolution of hydro-osmotic organs and neurohypophysial hormones in amphibia

Amphibia have realized the transition from water to land and their physiology has characteristics of both aquatic and terrestrial vertebrates. Whereas the tadpole has the osmoregulatory status of a FW bony fish, involving gut, kidney and gills, a new developmental program is set up during metamorphosis and the adult acquires a partially terrestrial osmoregulatory system. The skin becomes permeable to gas and water so that the animal breathes mainly through the skin and compensates the strong evaporative water loss by cutaneous water uptake (see Uchiyama and Yoshizawa this volume). In addition, water filtered through the glomerulus is in part reabsorbed from the nephron and from the urinary bladder which functions in water storage. The nephron has not yet developed the Henle's loop as in birds and mammals so that urine produced is hypotonic to plasma. The water permeability of the three organs, nephron, bladder and skin is regulated by the neurohypophysial hormones vasotocin and hydrins (see below).

6.1 *The neurohypophysial hormones of amphibia*

The class of Amphibia comprises three orders: Apods, Urodela (tailed amphibians) and Anura. Chemical characterization of neurohypophysial hormones has been performed to date only in species of the third order. However chromatographic and pharmacological data suggest that the two main peptides vasotocin ([Ile^3]vasopressin) and mesotocin ([Ile^8]-oxytocin) that are present in all non-mammalian tetrapods, occur also in Urodela. All neurohypophysial hormones derive from protein precursors that are processed into active nonapeptides and neurophysins in hypothalamic secretory neurons and are stored in secretory granules (Michel *et al.*, 1987).

Vasotocin. Vasotocin ([Ile^3]-vasopressin) (*Figure 3*) is an alpha-amidated nonapeptide that replaces mammalian vasopressin in all non-mammalian vertebrates. The substitution of the aliphatic hydrophobic residue isoleucine for the aromatic and hydrophobic phenylalanine in position 3, modifies the pharmacological properties and therefore the affinities to specific receptors. Vasotocin displays an antidiuretic activity of 260 I.U. μmol^{-1} measured in a rat bioassay compared to 460 I.U μmol^{-1} for vasopressin (Berde and Boissonnas, 1968). However in contrast to vasopressin, it is also active on the oxytocin receptor of rat uterus: 125 I.U. μmol^{-1}, whereas vasopressin displays 17 I.U. $fmol^{-1}$. Because vasotocin is the only neurohypophysial peptide found in cyclostomes, the most primitive vertebrates, it is considered the 'ancestral' molecule that may have appeared first some 450 million years ago. Duplication of the provasotocin gene during the subsequent rise of the fish led to two neurohypophysial peptide lineages that can be traced along the bony vertebrate scale (Acher, 1995).

Mesotocin. Mesotocin ([Ile^8]-oxytocin) (*Figure 3*) is the second peptide found in the neurohypophysis of non-mammalian tetrapods (Acher, 1995). It differs from mammalian oxytocin by the substitution of the aliphatic hydrophobic residue isoleucin for the aliphatic hydrophobic leucin in position 8. This substitution does not qualitatively modify the pharmacological profile: mesotocin displays an oxytocic activity of 291 I.U. μmol^{-1} on rat uterus compared with 450 I.U. μmol^{-1} for oxytocin, the rat endogenous hormone (Berde and Boissonnas, 1968). Mesotocin is found in all non-mammalian tetrapods from lungfish to amphibians, reptiles and birds and also in most marsupials. It is of interest to mention that in the African toad *Bufo regularis*, that survives in an arid environments, mesotocin is replaced by seritocin ([Ser^5,Ile^8]-oxytocin or [Ser^5]-mesotocin). The pharmacological and physiological properties of this peptide are still unknown.

Hydrins. Isolation of neurohypophysial peptides by high-pressure liquid chromatography (HPLC) from a number of anuran Amphibia has revealed the presence, next to vasotocin and mesotocin, of a peptide that is not detected in other non-mammalian tetrapods (Rouillé *et al.*, 1989). In a dozen frogs and toads from Europe, North and South America, Africa and Asia, this peptide has been identified as vasotocin-Gly (hydrin 2) by amino acid sequence and co-elution with the synthetic counterpart in HPLC analysis. In *Xenopus laevis,* this additional peptide is vasotocinyl-Gly-Lys-Arg (hydrin 1). Both result from a down-regulation in the processing of provasotocin in the neurosecretory granule, hydrin 1 by a decrease in the activity of carboxypeptidase E or H, hydrin 2 by the lowering of the polypeptidylglycine-alpha-amidating enzymatic (PAM) system activity. Aside from an extension in the polypeptide length when compared with vasotocin (dodecapeptide and decapeptide instead of nonapeptide), the most important difference lies in the nature of the C-terminal group: an uncharged carboxamide in vasotocin, an anionic carboxylic group in hydrins. As a result, the conformations of the C-terminal parts are different in the respective molecules and therefore the reaction with receptors can be modified. Vasotocin and hydrins possess the same ring as oxytocin but are about 40 times more active than oxytocin on effecting skin or bladder water permeation.

Hydrins appear to be absent from Urodela such as *Ambystoma*. In addition, it was previously recognized that vasotocin is virtually inactive in the control of water permeation of skin of urodeles.

6.2 Molecular and physiological functions of vasotocin

Molecular functions. The interaction of vasotocin with receptors may be predicted by analogy to those of vasopressin established in mammals. Two main types of G protein-coupled vasopressin receptors are known: a V1-type that activates phospholipase Cβ, that in turn hydrolyses phosphoinositides into diacylglycerol and inositoltrisphosphate; the latter triggers an increase of intra-cellular Ca^{2+} and subsequently an increase of protein kinase C activity; the V2-type stimulates adenylate cyclase and increases PKA activity via cAMP.

These two types of receptors have been recognized for vasotocin in amphibians by detecting their specific second messengers after hormone stimulation. They have been identified in the nephron of the bullfrog kidney (Uchiyama, 1994) and *R. ridibunda* kidney (Ammar *et al.*, 1995). A vasotocin V1-receptor from the teleostean white sucker (*Catastomus commersonii*) and a vasotocin receptor from *X. laevis* has been partially sequenced by molecular cloning (Mahlman *et al.*, 1994). These first receptors of lower vertebrates show a high level of amino acid sequence identity with mammalian vasopressin receptors, suggesting strongly the occurrence of two lineages V1a and V2 receptors throughout the development of bony vertebrates.

A mesotocin receptor from *B. marinus* urinary bladder has been cloned that exhibits 70% sequence identity with the mammalian oxytocin receptor (Akhundova *et al.*, 1996); an isotocin receptor from *Catastomus commersonii* displays 66% identity with the amphibian mesotocin receptor (Haussmann *et al.*, 1995). Clearly, both a vasotocin/vasopressin-receptor and an isotocin/mesotocin/oxytocin-receptor lineage co-evolved in parallel. The highly conserved structure of the neurohypophysial hormones during the evolution of the bony vertebrates is matched by a similar high structural stability of their receptors.

Physiology. Vasotocin acts on the three osmoregulatory organs of amphibia (skin, bladder and nephron) by increasing the number of water channels as does vasopressin on the collecting duct principal cells of mammals.

In anuran Amphibia, the nephron, the urinary bladder and the skin are the sites of water (re)absorption and these organs respond to vasotocin by increasing their permeability to water. It is quite possible that three different but related vasotocin V2-receptors are involved in the three organs. Concerning amphibian aquaporins, a member of the MIP-superfamily has been cloned using cDNA from frog (*R. esculenta*) urinary bladder (Abrami *et al.*, 1994). This frog aquaporin FA-CHIP (or fAQP1), displays 77.4% identity with human AQP1 and only 42.4% with human vasopressin-regulated AQP2. So, identification of frog AQPs 2 and 3 awaits further study.

The regulatory role of vasotocin in amphibians (see also Uchiyama and Yoshizawa in this volume) is substantiated by a three- to four-fold increase of its plasma titre when the euryhaline frog *R. cancrivora* is dehydrated (weight loss 4–10%) by transfer from FW to saline (400–600 mOsm kg^{-1} NaCl for 3–4 hr). Vasotocin promotes cutaneous water uptake in a variety of anurans except *Xenopus laevis* and, on the other hand, increases the permeability to water and urea of the frog urinary bladder (Bentley, 1971). The action appears mediated through a V2-type receptor in both the urinary bladder and the skin.

6.3 *Putative molecular and physiological functions of hydrins*

The specific adaptation of anuran Amphibia, that process provasotocin in the neurohypophysial secretory granules to two hormones (vasotocin and hydrin), suggests an important role for the hydrin in the particular physiology of the amphibian compared to that of the other vasotocin-bearing vertebrates. Because of its strong structural similarity with vasotocin, hydrin 2 (vasotocinyl-Gly) could act on some vasotocin receptors. However the C-terminal amidated group of neurohypophysial nonapeptide hormones, missing in hydrins, is essential for their biologial properties and therefore for their binding to their receptors.

Vasotocin, hydrin 1 and hydrin 2 have been compared for their activities on the three frog osmoregulatory organs, namely on kidney (antidiuretic activity), frog bladder permeation *in vitro* (gravimetric method) and frog skin permeation *in vivo* (body weight increase). It appears that hydrins are as active as vasotocin on the bladder, that hydrin 2 is more active than vasotocin on the skin, but that both hydrins are virtually devoid of antidiuretic activity in frog. Furthermore when assayed on mammalian organs, hydrins have weak uterotonic (rat uterus contraction), pressor (rat blood pressure) and antidiuretic (rat antidiuresis) activities in contrast to vasotocin.

We assume that hydrins have weak affinities for vasopressin and oxytocin receptors in mammals in contrast to vasotocin because of their charged C-terminal anionic group. In frog, it is possible that the vasotocin receptor of the kidney is different from vasotocin receptors of the bladder and the skin and that hydrins can act on the latter. But another interpretation would be that hydrins possess their own receptors in bladder and skin distinct from those for vasotocin and that hydrin receptors are missing in the kidney. So the adaptive evolution would have build up in amphibians a complementary rehydration function based on the regulated water permeability of the skin and the bladder in order to compensate the considerable water loss through evaporation. Whereas vasotocin seems mainly aimed for renal reabsorption of internal water, a function preserved in all non-mammalian tetrapods, hydrin 2 would be devoted to the uptake of environmental water (Acher *et al.*, 1997b). This proposed new function has received little attention so far since hydrin 2 derives from the same precursor as vasotocin and because its putative skin and bladder receptors probably have the same pattern as that of neurohypophysial hormone receptors, arising from copies of vasotocin receptor genes. The physiological function of mesotocin and of its specific bladder receptor in water homeostasis of Amphibia remains to be explored.

References

Abrami, L., Simon, M., Rousselet, G., Berthonaud, V., Buhler, J.M. and Ripoche, P. (1994) Sequence and functional expression of an amphibian water channel, FA-CHIP: A new member of the MIP family. *Biochim. Biophys. Acta* **1192**: 147–151.

Acher, R. (1995) Evolution of neurohypophysial control of water homeostasis: integrative biology of molecular, cellular and organismal aspects. In: *Neurohypophysis: Recent Progress of Vasopressin and Oxytocin Research* (eds T. Saito, K. Kurokawa, and S. Yoshida). Elsevier Science, pp 39–54.

Acher, R. (1996) Molecular evolution of fish neurohypophysial hormones: Neutral and selective evolutionary mechanisms. *Gen. Comp. Endocrinol.* **102:** 157–172.

Acher, R. and Chauvet, J. (1995) The neurohypophysial endocrine regulatory cascade: precursors, mediators, receptors, effectors. *Front. Neuroendocrinol.* **16:** 237–289.

Acher, R., Chauvet, J., Chauvet, M-T., Michel, G. and Rouillé, Y. (1997a) Molecular evolution of neurohypophysial hormones in relation to osmoregulation: the two fish options. *Fish. Physiol. Biochem.* **17**: 325–332.

Acher, R., Chauvet, J. and Rouillé, Y. (1997b) Adaptive evolution of water homeostasis regulation in amphibians: Vasotocin and hydrins. *Biology of the Cell* **89**: 283–291.

Acher, R., Chauvet, J., Chauvet, M-T. and Rouillé, Y. (1999) Unique evolution of neurohypophysial hormones in cartilaginous fishes: possible implications for urea-based osmoregulation. *J. Exp. Zool.* **284:** 475–484.

Akhundova, A., Getmonova, E., Gorbulev, V., Carnazzi, E., Eggena, P. and Fahrenholz, F. (1996) Cloning and functional characterization of the amphibian mesotocin receptor, a member of the oxytocin/vasopressin receptor superfamily. *Eur. J. Biochem.* **237:** 759–767.

Ammar, A., Roseau, S. and Butlen, D. (1995) Vasotocin-sensitive adenylate cyclase in frog glomeruli. *Gen. Comp. Endocrinol.* **98:** 102–110.

Bentley, P.J. (1971) *Endocrines and Osmoregulation, Zoology and Ecology*, Vol. 1. Springer, Berlin.

Bentley, P.J. (1972) Comparative endocrinology and osmoregulation. *Fed. Proc.* **31**: 1583–1586.

Berde, B. and Boissonnas, R.A. (1968) Basic pharmacological properties of synthetic analogues and homologues of the neurohypophysial hormones. *Handbk Exp. Pharmacol.* **23**: 802–870.

Bourque, C.W., Oliet, S.H.R. and Richard, D. (1994) Osmoreceptors, osmoreception, and osmoregulation. *Front. Neuroendocrinol.* **5**: 231–274.

Chauvet, J., Rouillé, Y., Chauveau, C., Chauvet, M-T. and Acher, R. (1994) Special evolution of neurohypophysial hormones in cartilaginous fishes: Asvatocin and phasvatocin, two new oxytocin-like peptides in the spotted dogfish *(Scyliorhinus caniculus). Proc. Natl Acad. Sci. USA* **91**: 11266–11270.

Doolittle, R.F., Fen, D-F., Tsang, S., Cho, G. and Little, E. (1996) Determining divergence times of the major kingdoms of living organisms with a protein clock. *Science* **271**: 470–477.

Evans, D.H. (1993) Osmotic and ionic regulation. In: *The Physiology of Fishes* (ed. D.H. Evans). CRC Press, Boca Raton, FL, pp 315–341.

Fernandez-Chacon, R. and Südhof, T.C. (1999) Genetics of synaptic vesicle function: Toward the complete functional anatomy of an organelle. *Ann. Rev. Physiol.* **61:** 753–776.

Hara, Y., Battey, J. and Gainer, H. (1990) Structure of mouse vasopressin and oxytocin genes. *Mol. Brain Res.* **8:** 319–324.

Hausmann, H., Meyerhof, W., Zwiers, H., Lederis, K. and Richter, D. (1995) Teleost isotocin receptor: structure, functional expression, mRNA distribution and phylogeny. *FEBS Lett.* **370:** 227–230.

Ivell, R. and Richter, D. (1984) Structure and comparison of the oxytocin and vasopressin genes of rat. *Proc. Natl Acad. Sci. USA* **81:** 2006–2010.

Katz, U. (1990) Osmoregulatory mechanisms in the adaptation of amphibia to high salt solutions. In: *Animal Nutrition and Transport Processes 2. Transport, Respiration and Excretion: Comparative and Environmental Aspects.* Comp. Physiol. Vol. 6 (eds J.P. Truchot and B. Lahlou). Karger, Basel, pp. 26–38.

Klussmann, E., Maric, K. and Rosenthal, W. (2000) The mechanisms of Aquaporin control in the renal collecting duct. In: *Water Transport Across Biological Membranes. Physiology, Biochemistry and Pharmacology,* Vol. 141 (eds R. Greger and W. Rosenthal). Springer, Berlin.

Kültz, D. and Csonka, L. (1999) What sets the TonE during osmotic stress? *Proc. Natl Acad. Sci. USA* **96:** 1814–1816.

Lacy, E.R. and Reale, E. (1995) Functional morphology of the elasmobranch nephron and retention of urea. In: *Cellular and Molecular Approaches to Fish Ionic Regulation* (eds C.M. Wood and T.J. Shuttleworth). Academic Press, San Diego, CA.

Mahlman, S., Meyerhof, W., Hausmann, H., Heierhorst, J., Schönrock, C., Zwiers, H., Lederis, K. and Richter, D. (1994) Structure, function and phylogeny of [Arg^8]-vasotocin receptor from teleost fish and toad. *Proc. Natl Acad. Sci USA* **91:** 1342–1345.

Matalon, S. (1999) Sodium channels in alveolar epithelial cells: Molecular characterization, biophysical properties, and physiological significance. *Ann. Rev. Physiol.* **61:** 627–661.

Michel, G., Chauvet, J., Chauvet, M.T. and Acher, R. (1987) One-step processing of the amphibian vasotocin precursor: structure of a frog (*Rana esculenta*) 'big' neurophysin. *Biochem. Biophys. Res. Commun.* **149**: 538–544.

Miyakawa, H., Woo, S.K., Dahl, S.C., Handler, J.S. and Kwon, H.M. (1999) Tonicity-responsive enhancer binding protein, a Rel-like protein that stimulates transcription in response to hypertonicity. *Proc. Natl Acad. Sci. USA* **96**: 2538–2542.

Mojzsis, S.J., Arrhenius, G., McKeegan, K.D., Harrison, T.M., Nutman, A.P. and Friend, C.R.L. (1996) Evidence for life on Earth before 3,800 million years ago. *Nature* **384**: 55–59.

Nelson, J.S. (1994) *Fishes of the World* , 3rd edn. John Wiley, New York.

Perlman, D.F. and Goldstein, L. (1988) Nitrogen metabolism. In: *Physiology of Elasmobranch Fishes* (ed. T.J. Shuttleworth). Springer, Berlin.

Quan, A.H. and Cogan, M.G. (1993) Body fluid compartment and water balance. In: *Clinical Disturbances of Water Metabolism* (eds D.W. Seldin and G. Giebisch). Raven Press, New York, pp. 1–9.

Rosing, M.T. (1999) ^{13}C-depleted carbon microparticles in > 3700-Ma sea-floor sedimentary rocks from West Greenland. *Science* **283:** 674–676.

Rouillé, Y., Michel, G., Chauvet, M-T., Chauvet, J. and Acher, R. (1989) Hydrins, hydro-osmotic neurohypophysial peptides: Osmoregulatory adaptation in amphibians through vasotocin precursor processing. *Proc. Natl Acad. Sci. USA* **86:** 5272–5275.

Ruppert, S., Scherer, G. and Schütz, G. (1984) Recent gene conversion involving bovine vasopressin and oxytocin precursor genes suggested by nucleotide sequence. *Nature* **308:** 554–557.

Sausville, E., Carney, D. and Battey, J. (1985) The human vasopressin gene is linked to the oxytocin gene and is selectively expressed in a cultured lung cancer cell line. *J. Biol. Chem.* **260**: 10236–10241.

Schopf, J.W. (1993) Microfossils of the Early Archean Apex Chert: New evidence of the antiquity of life. *Science* **260:** 640–646.

Setou, M., Nakagawa, T., Seog, D.H. and Hirokawa, N. (2000) Kinesin superfamily motor protein KIF17 and mLin-10 in NMDA receptor-containing vesicle transport. *Science* **288**: 1796–1802.

Shayakul, C., Steel, A. and Hediger, S. (1996) Molecular cloning and characterization of the vasopressin-regulated urea transporter of rat kidney collecting duct. *J. Clin. Invest.* **98**: 2580–2587.

Shuttleworth, T.J. (1988) Salt and water balance – extrarenal mechanisms. In: *Physiology of Elasmobranch Fishes* (ed. T.J. Shuttleworth). Springer, Berlin.

Thrasher, T.N. (1994) Baroreceptor regulation of vasopressin and renin secretion: low-pressure versus high-pressure receptors. *Front. Neuroendocrinol.* **15**: 157–196.

Uchiyama, M. (1994) Sites of action of arginine vasotocin in the nephron of the bullfrog kidney. *Gen. Comp. Endocrinol.* **94**: 366–373.

Van der Heide, T. and Poolman, B. (2000) Osmoregulated ABC-transport system of *Lactococcus lactis* senses water stress via changes in the physical state of the membrane. *Proc. Natl Acad. Sci. USA* **97**: 7102–7106.

Wang, W., Hebert, S.C. and Giebisch, G. (1997) Renal K^+ channels: Structure and function. *Ann. Rev. Physiol.* **59**: 413–436.

Index